AN
INTRODUCTION TO
MICROBIOLOGY

PHARMACEUTICAL MONOGRAPHS

GENERAL EDITOR

J. B. STENLAKE, D.Sc., Ph.D., F.P.S., F.R.I.C., F.R.S.E.

Professor of Pharmacy, University of Strathclyde

Volume 1

AN INTRODUCTION TO MICROBIOLOGY

W. B. Hugo B.Pharm., Ph.D. (Lond), F.P.S.

Volume 2

AN INTRODUCTION TO PARASITOLOGY

John M. Watson D.Sc. (Lond), F.P.S.

Volume 3

STERILISATION AND DISINFECTION

T. D. Whittet B.Sc., Ph.D., F.P.S., F.R.I.C., D.B.A.
W. B. Hugo B.Pharm., Ph.D. (Lond), F.P.S.
G. R. Wilkinson F.P.S.

Volume 4

POLYSACCHARIDES, PEPTIDES AND PROTEINS

R. T. Coutts B.Sc., Ph.D., A.R.C.S.T., M.P.S.
G. A. Smail B.Sc., A.R.C.S.T., M.P.S.

Volume 5

IMMUNOLOGICAL AND BLOOD PRODUCTS

Mary Dawson Ph.D., F.P.S.
G. R. Milne F.P.S.

Volume 6

SURGICAL DRESSINGS, LIGATURES AND SUTURES

F. Fish B.Pharm., Ph.D., F.P.S.
J. Owen Dawson B.Sc., F.P.S.

Volume 7

UNIT PROCESSES IN PHARMACY

David Ganderton B.Pharm., Ph.D.

Volume 8

TABLETS AND TABLETTING

H. Burlinson F.P.S. D.B.A.

Volume 9

PHARMACEUTICAL AND COSMETIC PRODUCTS FOR TOPICAL ADMINISTRATION

N. J. Van Abbe F.P.S. M.Inst.Pkg.
R. I. C. Spearman Ph.D.
A. Jarrett M.B., D.Sc., M.R.C.Path., F.R.C.P.

AN INTRODUCTION TO MICROBIOLOGY

BY

W. B. HUGO

B.Pharm., Ph.D. (Lond), F.P.S.

University of Nottingham

Second Edition

WILLIAM HEINEMANN MEDICAL BOOKS LTD
LONDON

First published 1964
Second edition 1972

ISBN 0 433 15601 5

Printed in Great Britain by
Butler & Tanner Ltd, Frome and London

GENERAL PREFACE

The aim of this series of pharmaceutical monographs is to provide an up-to-date series of short publications for teaching general and specialised topics to undergraduate students of pharmacy and allied subjects. Each monograph in the series is the work of an expert or group of experts actively engaged in teaching or practice. A few of the introductory monographs are sufficiently comprehensive to warrant publication as individual volumes, but for convenience some of the shorter monographs on related subjects have been collected together for publication under one cover.

Each monograph is intended to serve as the basis for a group of lectures or tutorials in the honours and pre-honours years of undergraduate courses in pharmacy and allied subjects in British and Commonwealth Universities and, of necessity, some monographs are slanted towards the more specific requirements of these countries. We have, however, endeavoured to keep the monographs on a general plane to ensure their suitability for use in other parts of the world.

An attempt has been made to present the subject matter of individual monographs in such detail that it provides a permanent record for study purposes capable of being used by students in lieu of lecture notes. Each monograph, however, sets out to provide not merely a detailed account of essential subject matter, such as would be required for examination purposes, but also seeks to indicate its relevance and importance to pharmaceutical studies in general. In this respect, the more advanced monographs extend naturally to the boundaries of knowledge in all major aspects, and wherever possible present appropriate rival views and hypotheses in sufficient detail for the student to grasp their essential detail without reference to the original. The texts are, however, referenced to provide additional sources of information.

I am indebted to the authors of the individual monographs for their willingness to collaborate with me in the preparation of this series. I should also like to express my thanks to my colleagues and many friends for their help and advice in framing the series and for discussions on individual monographs. I should further like to express my sincere thanks to Mrs S. Cohen for invaluable secretarial assistance.

J. B. S.

PREFACE TO THE FIRST EDITION

The treatment of bacteriology in this introduction to the subject has been chosen quite deliberately as representing the most interesting approach. To avoid plunging the student into a maelstrom of names and species with the details by which they are distinguished, the fascinating picture of the overall anatomy of the cell, with brief references to methods of study, has formed the opening chapter. Thereafter the behaviour of cells during growth and the factors affecting growth are considered and their relevance to applied aspects of microbiology indicated. There then follows a chapter on metabolism, an aspect of microbiology often ignored or too scantily treated in pharmacy courses. Finally a compilation of microbial species arranged according to current classification schemes is presented.

The introduction exceeds the length originally allocated by the Scientific Editor and the author apologises if readers feel that certain sections have still not received the detailed treatment they would wish. Nevertheless the priorities and treatment have endeavoured to highlight the facets of microbiology of importance to a modern student of pharmacy, for if pharmacy is to stand firmly as an important technology in its own right the days of the empirical approach to the subject must be numbered.

References have been made at the end of each chapter to detailed monographs and texts which themselves contain many further references. Only one or two original papers of particular relevance to the text have been cited.

The author acknowledges with thanks the help he received from his wife who read the whole of the manuscript and made valuable suggestions. Likewise to Mr J. J. Lewis who kindly agreed to read the manuscript also and made many general comments as to style and points of presentation and to Dr Barbara Lund for help on specific points in the monograph and for reading the proofs.

The author thanks Mrs J. Crossland who prepared the typescript from a rather untidy manuscript.

Individual workers and pharmaceutical manufacturers are thanked for their help and provision of photographs and experimental data. Their several contributions are acknowledged in the text.

August 1964 W. B. H.

PREFACE TO THE SECOND EDITION

The second edition of this volume has been revised and extended and errors appearing in the first edition corrected.

Both the scientific editor and the publishers felt that radical extension of the subject matter and depth of coverage would destroy the intent of the series, of which this book forms the first volume, and with their views I have been happy to concur.

W. B. H.

CONTENTS

Chapter *Page*

1. The Bacterial Cell 1

2. Bacterial Growth and Factors Affecting Bacterial Growth 37

3. Bacterial Metabolism 71

4. General Properties of Micro-organisms of Significance in Pharmacy 117

Index 157

A NOTE ON THE NOMENCLATURE OF UNITS OF LENGTH AND OF THE KETO (OXO) ACIDS

Since the first edition of this book was prepared two units of length have changed their name and symbol. As students will find the old nomenclature in texts and the literature the following will help to explain the change.

The mμ (millimu) becomes the nm (nanometre); these units equal 10^{-9} metre; the μ (mu) becomes the μm (mumetre); and these units represent 10^{-6} metre.

Keto acids are now usually called oxo acids and the greek letter indicating the positioning of the group is replaced by a roman numeral, thus α-ketoglutaric acid becomes 2-oxoglutaric acid and β-ketoadipic acid becomes 3-oxoadipic acid.

CHAPTER 1

THE BACTERIAL CELL

The present detailed knowledge of the anatomy of the bacterial cell represents the cumulation of a century of research. Early information was gathered by careful microscopic study using light (as distinct from electron) microscopy and a variety of special staining techniques. One of these, discovered in 1884 by Gram (page 29), divided bacteria into two groups dependent on whether they remained stained (Gram-positive) or not (Gram-negative) on washing with ethanol. This difference was later found to be linked with other general differences in properties within the two groups.

More recently the study of bacteria by electron microscopy has confirmed and extended the findings of the early workers. It is now possible to cut sections of bacteria thin enough for examination by electron microscopy, while scanning electron microscopy is playing an increasingly important role in the study of microbial anatomy.

In addition much information has been gained by the fragmentation of bacteria by mechanical or chemical methods followed by separation of components by differential centrifugation. These components have been subsequently examined in the electron and scanning electron microscope and also subjected to chemical analysis.

SHAPE AND MORPHOLOGY

The bacteria are unicellular organisms and it is important when considering bacterial cells to gain first an impression of their shape and size. The true bacteria exist in two main morphological variants, as rod-shaped cells with rounded or sometimes pointed ends, and as spherical or almost spherical cells called cocci.

Four variants of the rod-shaped form are found: rods with one convolution (*Vibrio*), rods with several convolutions (*Spirillum*), rods which tend to form chains with the cells arranged end to end (some lactobacilli) and those that may show branched forms as seen amongst some of the Actinomycetales. Variants of the coccoid form depend on the degree and manner of aggregation of the cells

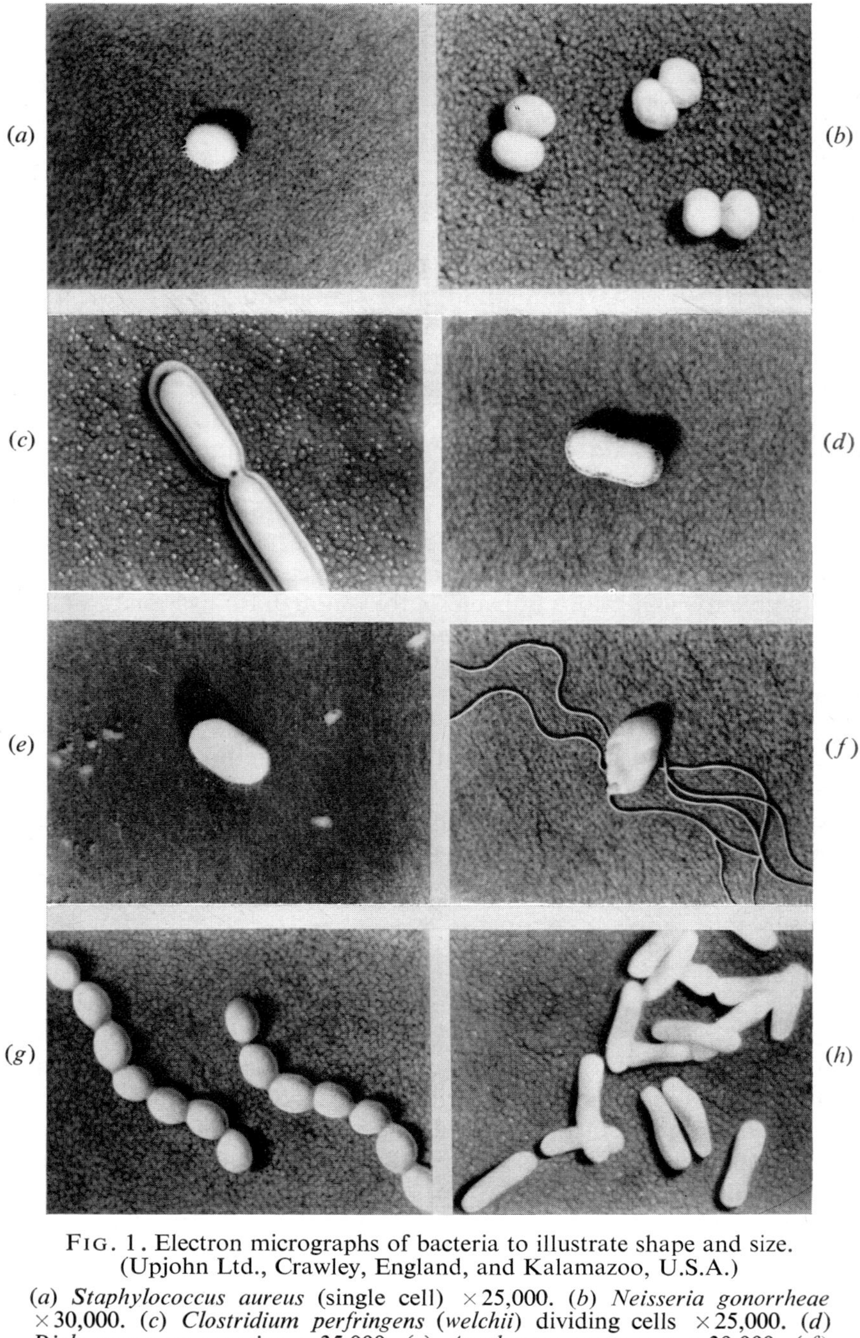

FIG. 1. Electron micrographs of bacteria to illustrate shape and size. (Upjohn Ltd., Crawley, England, and Kalamazoo, U.S.A.)

(*a*) *Staphylococcus aureus* (single cell) ×25,000. (*b*) *Neisseria gonorrheae* ×30,000. (*c*) *Clostridium perfringens* (*welchii*) dividing cells ×25,000. (*d*) *Diplococcus pneumoniae* ×35,000. (*e*) *Aerobacter aerogenes* ×30,000. (*f*) *Salmonella schottmülleri* (*paratyphi*) ×25,000. (*g*) *Streptococcus viridans* ×30,000. (*h*) *Corynebacterium diphtheriae* ×25,000.

after division during reproduction. Coccoid forms may occur singly, as pairs (diplococci), as cubical aggregates of the sphere (sarcinae), in irregular clusters, often likened to a bunch of grapes (staphylococci) or in chains (streptococci). Amongst some species, for example *Leuconostoc*, ovoid cells are found. Much use has been made of cellular morphology in evolving systems of bacterial classification (Chapter 4). Micrographs of bacteria are shown in Figs. 1 and 2.

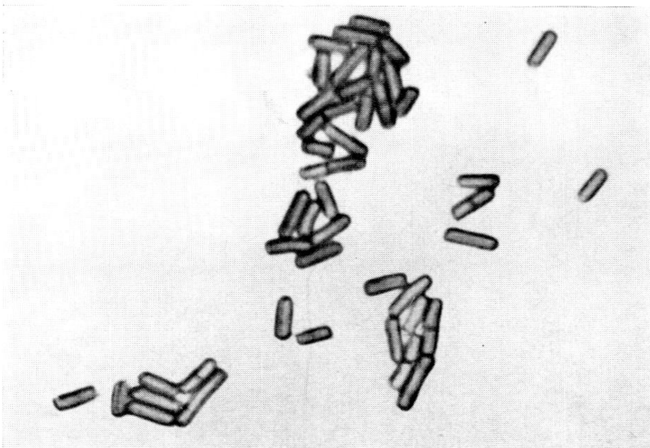

FIG. 2. *Clostridium perfringens* (*welchii*), ×2700. (Glaxo Laboratories, England)

The unit of size in measuring bacteria is the micrometer, μm (1 μm = $\frac{1}{1000}$ mm). *Escherichia coli*, a typical rod, may be 3 μm in length and 0·75 μm in diameter, while *Staphylococcus aureus*, a coccoid form, may be 0·75–1·0 μm in diameter.

BACTERIAL ANATOMY

The main features of the anatomy of the bacterial cell are shown diagrammatically in Fig. 3.

The Cell Wall

Structures immediately exterior to the cytoplasmic membrane may be defined as the cell wall. The wall is a rigid structure, which is responsible for the characteristic shape of the bacterial cell and withstands an internal osmotic pressure which may range from 6 to 20 atmospheres according to the species. It acts as a mechanical barrier to the entry of some large molecules into the cell and in the case of *Staphylococcus aureus*, for example, is thought to

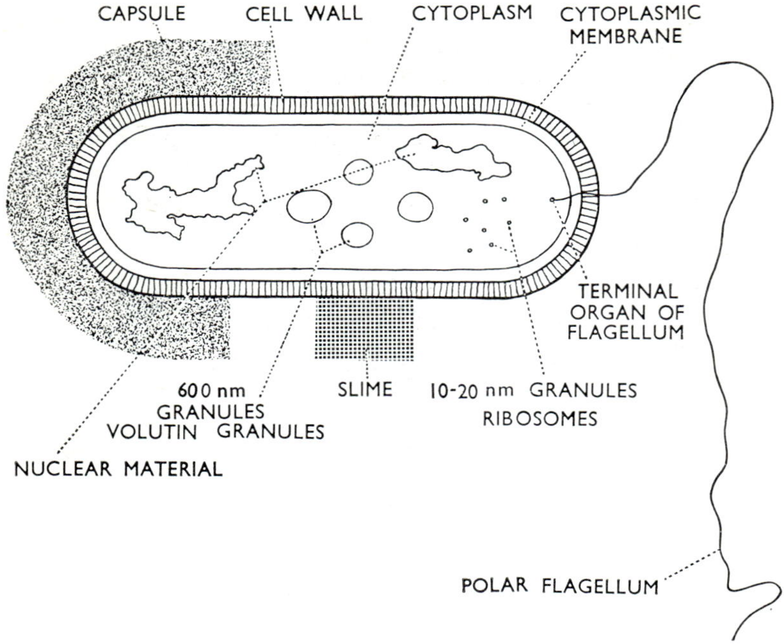

FIG. 3. Diagram to illustrate the main features of the bacterial cell

have an effective pore diameter of the order of 1 nm (1 nm $= \frac{1}{1000}\ \mu m$). The mechanism by which large molecules move in and out of the bacterial cells is at the moment controversial and molecular size is certainly not the only criterion controlling the movement of molecules and ions in and out of the cell.

Detailed studies of the chemical structures involved in the cell wall became possible following the publication in 1948 of a method of preparing bacterial cell walls reasonably free from other cell components. This was achieved by mixing a bacterial suspension with glass beads of about 0·07 mm diameter and agitating rapidly; bacterial cells are ruptured by this process; washing and careful centrifugation resulted in a preparation rich in cell wall material. A photograph of such a cell wall preparation of *Staphylococcus aureus* taken with an electron microscope is shown in Fig. 4.

Wall structure varies considerably between bacterial species and these structures play important roles both in determining the fundamental properties of the organism and in determining their response to drug action.

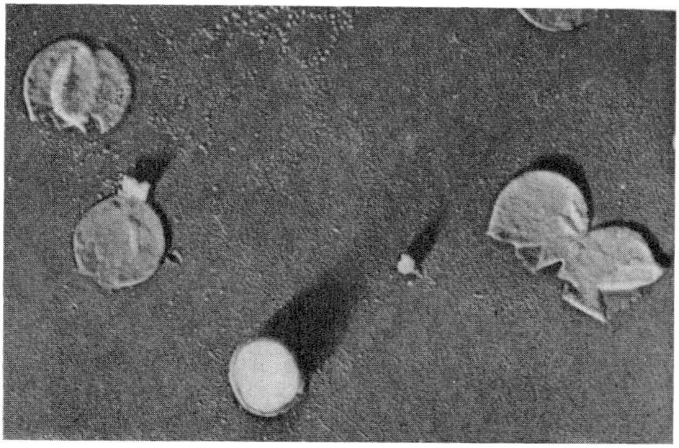

FIG. 4. Cells of *Staphylococcus aureus* disintegrated by shaking with small glass beads. One cell in the group has remained intact. (Thanks are due to Dr A. S. McFarlane and the editor of the *British Medical Journal* for permission to publish and Burroughs Wellcome and Co. for providing the original photograph)

Before 1948 very little was known of the chemical structure of the bacterial cell wall but since then a large literature has accumulated.

A fundamental structure found in the cell walls of all bacteria and the blue-green algae i.e. procaryotic cells (page 118) has been called mucopeptide, murein, glycosaminopeptide, glycopeptide, peptidoglycan or the murein sacculus. Peptidoglycan is the preferred term.

The first hint as to the nature of this component came in 1952 when Park, while studying the action of penicillin on staphylococci, noted the accumulation of a complex nucleotide which contained what was later to be confirmed as a component of the cell wall.

As stated above it was known that, like other cells, a strong or rigid component must be present in bacterial cell walls to withstand the osmotic pressure exerted by the cytoplasmic contents of the cell and also to confer the characteristic shape of sphere or rod. Early observers notably Gardener in 1940 and Duguid in 1945 had observed that penicillin caused characteristic changes in growing bacterial cells which might be attributable to damage to, as Duguid put it, the outer supporting cell wall. These observations together with that of Park on the mode of action of penicillin thus provided the clues to the role and nature of this fundamental component. These observations were complimented by extensive analytical studies.

At this time, in a developing research field, considerable information is now available on cell wall structure and this will be considered below.

The peptidoglycan component. Nature has always used polymeric structures to confer rigidity in living organisms for example cellulose and modified celluloses are used thus in the plant kingdom and chitins in arthropods.

The peptidoglycan of bacteria is another such polymer possessing considerable structural rigidity. Chemically it consists of a linear polysaccharide or alternating β $(1 \rightarrow 4)$ linked units of N-acetyl glucosamine and N-acetyl-3-O-1-carboxylethylglucosamine. The latter component has the trivial name muramic acid. This linear polysaccharide is crossed linked by peptide chains the component amino acids of which vary both in type and method of linkage as between differing bacterial species. The whole cage-like wall structure is thus one large molecule and to this has been given the name mentioned above, the murein sacculus.

The antibiotics penicillin, cycloserine and vancomycin prevent, in different ways, the cross linking reaction in growing bacterial cultures i.e. in cultures in which new cell walls are being formed, thus damaging the newly forming cells which burst and lyse under the effect of the internal osmotic pressure of their cytoplasm, now no longer contained and protected by the weakened peptidoglycan component of the wall. It is interesting to note that the enzyme lysozyme, a component of egg white and lachrymal secretion, specifically hydrolyses the link between the two sugar components of the linear polymeric backbone of the wall i.e. between N-acetylglucosamine and N-acetyl-3-O-1-carboxyethylglucosamine, again with cocommitment lysin due to bursting.

A typical peptidoglycan, that of *Esch. coli* is shown diagrammatically in Fig. 5(a) and the detailed chemical structure of a repeating unit is shown in Fig. 5(b). The peptide moiety of *Staph. aureus* consists of L-alanine, D-glutamine, L-lysine and two molecules of D-alanine. In *Esch. coli* meso-diaminopimelic acid replaces the L-lysine of the *Staph. aureus* structure.

The final reaction which leads to the completed three dimensional and rigid wall peptidoglycan consists of the formation of cross-link between the peptide residues, in the linear polymers so

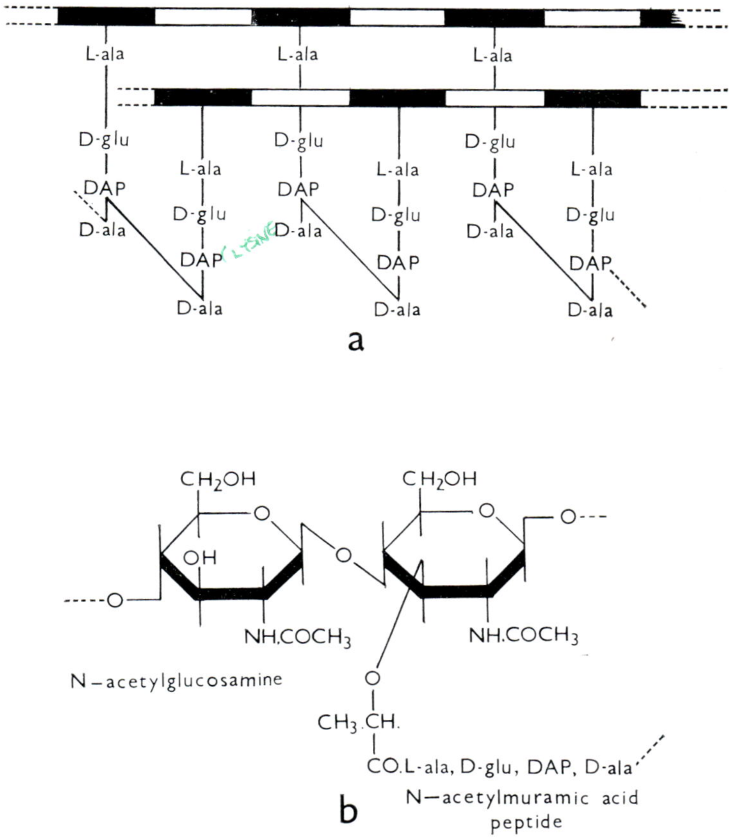

FIG. 5. (a) Peptidoglycan of *Esch. coli*. N-acetylmuramic acid, ▬ N-acety glucosamine ▭ (b) Repeating unit of peptidoglycan of *Esch. coli* (L-ala = L-alanine; D-glu = D-glutamine; DAP = diaminopimelic acid; D-ala = D-alanine.)

far described. This reaction involved the enzymatic removal of one of the terminal D-alanine molecules and the formation of a new peptide link with an adjacent lysine or diaminopimelic acid molecule.

In *Esch. coli* this occurs directly but in other organisms additional peptides are also inserted during the final stage of biosynthesis. In *Staph. aureus*, 5 glycine molecules appear in the final structure and other variants are known.

It is this final cross-linking which is affected by certain 'wall inhibiting' antibiotics, notably penicillin, vancomycin and ristocetin.

The structures of the amino acids mentioned above are depicted, with their standard abbreviations, in Table 1 (page 45).

Gram-positive walls. The differentiation of bacterial cells by staining into Gram-positive and Gram-negative has already been mentioned and a fundamental difference between these groups may be seen in the structure of their walls, however, the peptidoglycan component is common to all microbial cells and in Gram-positive organisms it may comprise 50–90 per cent of the total wall. Additional polymeric substances are found in Gram-positive cells and are characteristically absent from Gram-negative cells. They comprise a polyribitol or polyglycerol phosphate and have been given the general name of teichoic acids (Gk. teichos, wall) and may form 20–50 per cent of the wall. The amino acid alanine is also present, in techoic acids.

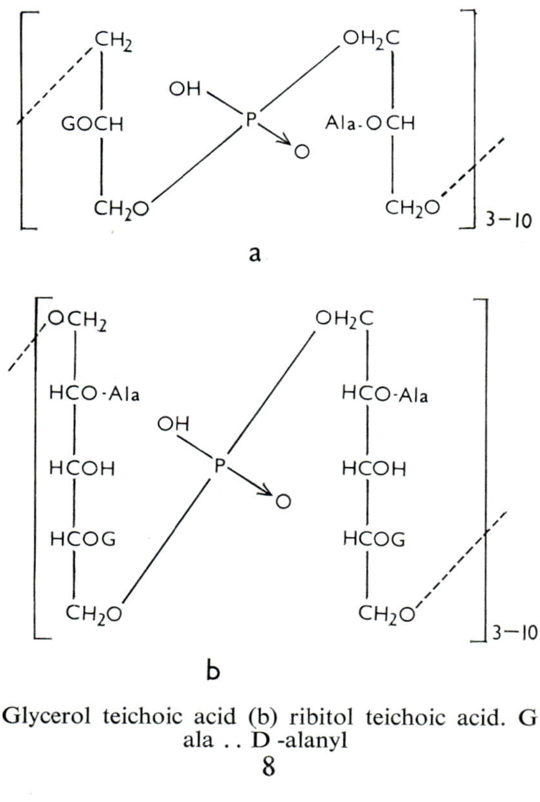

FIG. 6. (a) Glycerol teichoic acid (b) ribitol teichoic acid. G .. glycosyl; ala .. D -alanyl

Teichoic acids are attached to the peptidoglycan by a covalent link but do not conform rigidity. Being acid polysaccharides they are thought to function by collecting from the growth medium essential cations such as Mg^{++} important in enzyme activity (see page 74). They also form part of the antigenic structure of the cell (page 27).

Recently it has been shown that glycerol techoic acids occur in the bacterial membrane, as well as in cell walls.

The walls of the mycobacteria (page 139) which include the organism responsible for tuberculosis contain a high proportion of fatty material which include waxes.

Gram-negative cell walls. These are more complex than those of Gram-positive walls. Although they contain peptidoglycan it only forms some 3–12 per cent of the wall. The other components are lipoprotein (LP), lipopolysaccharide (LPS) and a globular protein. Teichoic acids are absent. The globular protein is thought to be attached by covalent links to the peptidoglycan while the lipoprotein and lipopolysaccharide occur in the outer layer possibly held in place by Mg^{++} ions. In some bacteria these two molecular species occur outside the globular protein as separate layers with the lipoprotein on the outside, in others, *Esch. coli*, for example, a mixed layer of lipoprotein and lipopolysaccharide is thought to exist (Fig. 7).

The LP/LPS behaves almost as an addition membrane affording some protection to the cell from antimicrobial agents for the concentration of many such substances required to inhibit cell growth is often higher in Gram-negative than Gram-positive bacteria; penicillin is a notable example. LP/LPS will also protect the peptidoglycan layer from digestion by lysozyme. LPS is an acid polysaccharide and is thought to function in the same way as do the teichoic acids of the Gram-positive cells, i.e. as collectors of cations. The LP/LPS is also associated with the toxins and pyrogenic (fever producing) substances in this group. (page 27)

The chemical structures and biosynthetic routes of the peptidoglycans, teichoic acids, and the lipopolysaccharides and lipoprotein of the bacterial cell wall are now known in great detail but research workers are still actively engaged in studying the walls of individual bacteria and many interesting variants of the structures of the well-studied organisms are appearing in the literature.

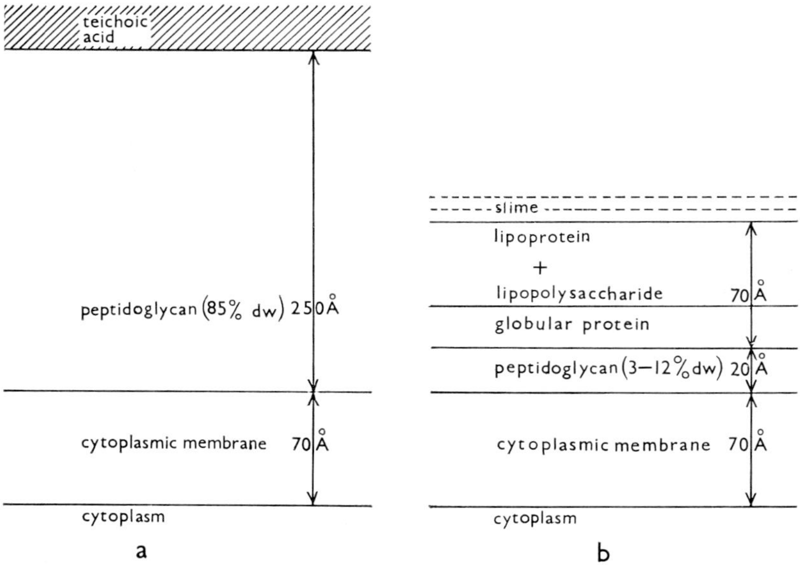

FIG. 7. The cell wall of (a) a Gram-positive bacterium and (b) a Gram-negative bacterium showing main differences in cell wall structure. 85 per cent dw or 3–12 per cent dw means that the peptidoglycan forms 85 or 3–12 per cent by weight of the total dry weight of the cells. 1 angstrom unit, Å, = 10nm = 10^{-10}m.

The Cytoplasmic Membrane

This non-rigid structure, sometimes called the osmotic barrier of the cell, acts as a semipermeable membrane and controls the passage of ions and molecules into and out of the cell.

Early evidence for the existence of a membrane which could act as an osmotic barrier was obtained at the beginning of this century when plasmolysis, that is shrinking of the cytoplasm away from the cell wall when bacteria were suddenly transferred to a medium of high osmotic pressure, was seen to occur. Furthermore, studies on the entry of fat-soluble material into the bacterial cell and the sensitivity of the cytoplasmic membrane to dissolution by fat solvents led to the hypothesis that this membrane was composed largely of fatty material.

Although mechanical disruption of bacterial cells has made possible the isolation of the cell wall and adherent cytoplasmic membrane free from cytoplasm, it has proved difficult in most instances to separate completely the cytoplasmic membrane itself by this technique alone. A method which has proved of value in the isolation and study of this membrane consists in preparing bacterial cells deficient in the rigid cell wall, either by digesting the cell wall with enzymes, or by growing cells in the presence of substances which prevent cell wall formation in new forming cells.

If these processes are carried out in normal culture media the high osmotic pressure within the cell results, once the wall is sufficiently weakened, in the immediate explosion of the cell; if, however, the digestion or inhibition of cell wall synthesis is carried out in medium containing a solute which will not pass into the cell, for example 0·33M sucrose to balance the internal osmotic pressure, then spherical bodies are formed irrespective of the original shape of the parent cell. If the chemical treatment results in the complete removal of the cell wall, the round body so formed is known as the cell protoplast. If only a rigid component of the cell wall is absent, as occurs with penicillin treatment, the round bodies are called spheroplasts. The spheroplast or protoplast has proved a valuable tool for studying permeability problems. By diluting the culture medium to allow explosion of the protoplast, membranes may then be separated for further study by centrifugation. Chemical analyses of these membrane preparations have confirmed histochemical evidence of their high lipid content and also demonstrated the presence of carbohydrate, i.e. the membrane glycerol techoic acids already referred to and protein, while the examination of their enzymic potentialities is yielding interesting results. Dehydrogenases, phosphatases and cytochrome oxidases are examples of enzymes identified with the cytoplasmic membrane. The role of these enzymes is considered in Chapter 3. In addition, the enzymes involved in the synthesis of the cell wall are now known to be associated with the membrane.

A group of special protein structures called permeases have also been demonstrated in bacterial membranes. Their existence was postulated on theoretical grounds in 1956–7 and they have now been isolated by allowing sucrose stabilised protoplasts to burst as already described. Permeases act as carriers of some of the metabolites needed by the cell from the exterior to the interior of the cell, and contribute to the general protein components of membranes.

Membrane structure. Chemical analyses of membranes had already shown that the main structural components were phospholipid and protein, and in 1935 Danielli and Davson proposed a three-layered structure for biological membranes in which they propose that a double layer of phospholipid molecules was sandwiched between a layer of protein (Fig. 8(a)). Although more

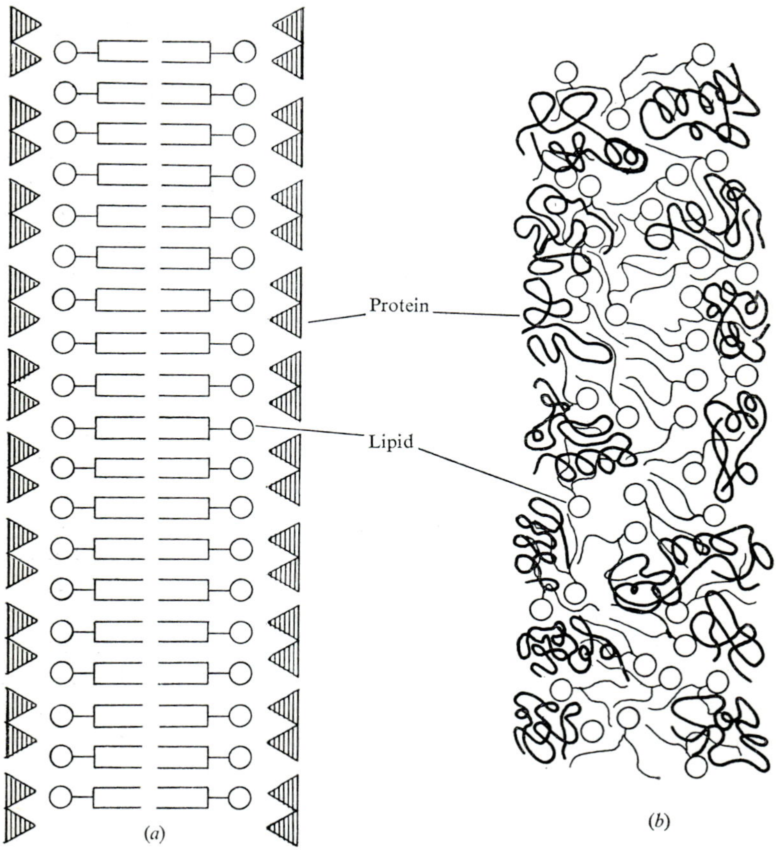

FIG. 8. Models of membrane structure. (See text.) (By permission of Professor E. F. Gale F.R.S. and the editors, *The Journal of General Microbiology*.)

detailed electron microscope studies together with biophysical observations at first gave no reason to doubt this structure, as experimentation became more sophisticated, it was found that an alternative structure was not only feasable but likely. The new structure proposed in 1962 by Luzzati and Husson was much less ordered than that of Danielli and Davson (Fig. 8(b)). This model can account for the existence of pores in the cytoplasmic membrane, the existence of which had been postulated on theoretical grounds by researchers in the field.

A sub-structure of the procaryotic membrane i.e. of bacteria and blue-green algae, is called a mesosome. Mesosomes are invaginations of the membrane. They show particularly well in the membranes of bacteria at the point of cell division. They may represent a device for the local concentration of wall synthesizing processes; they also contain accumulations of the cytochrome system, and have been likened, at least in function, to the mitochondrion of eukaryotic cells. Mesosomic structures have also been associated with the nuclear material of some bacteria and have again been thought to be prokaryotic equivalent of a nuclear membrane, page 118.

Mechanisms of transport. The transport of metabolites through the membrane may be divided into three categories. The first is called passive diffusion and in this process molecules and ions pass into the cell by a driving force which depends on either a concentration or charge difference. If permeases are also implicated in such a process it has been called facilitated diffusion.

The majority of molecules and ions enter the cell by a process known as active transport. Permeases may be involved and metabolic energy is necessary to carry the molecules and ions across the membrane. In the case of ion transport, electrical neutrality is maintained by the efflux from the cell of a corresponding number of ions bearing the same charge. As a result of the process of active transport a concentration within the cell of a molecule or ion may exceed, by a factor of 1000, the concentration outside the cell.

The Cytoplasm

Within the cytoplasmic membrane lies the cytoplasm. This generally has a granular structure and contains the nuclear material and all the metabolites and enzymes associated with the life process of the cell, other than those located in the cytoplasmic membrane. In those bacteria where detailed studies have so far been made, the major part of the cytoplasm has been shown to consist of granules 10–20 nm in diameter; these may be separated from crushed bacteria by careful centrifugation. Granules of the bacteria when viewed by electron microscopy. These granules contain ribonucleic acid and protein and are called ribosomes and are the site of protein synthesis. Structures of similar function have been found in animal and plant cells as well as in microorganisms. If amino acids containing radioactive atoms are supplied to growing cultures of bacteria, proteins containing these radioactive amino acids are found associated with the ribosome within 15 seconds; thereafter these labelled proteins appear as part of the soluble protein fraction of the bacterial cell (page 103). Enzymes involved in nucleic acid metabolism have also been isolated from ribosomes.

In addition to the 10–20 nm granules or ribosomes there exist also much larger granules which may be 600 nm in diameter which are particularly opaque in the electron microscope, and are consequently described as being electron-dense. Early bacteriologists observed these granules by conventional light microscopy and named them metachromatic granules because they caused a shift in the original colour, i.e. induced a spectral change in the dye used to display them. These electron dense bodies are also called volutin granules. Chemically they have been shown to contain polyphosphate compounds although these probably do not account for their total chemical complement.

Granular accumulations of lipid, often in the form of polyhydroxybutyric acid, and of carbohydrate have been reported in many bacterial species. Bacteria which are able to metabolise sulphur compounds and sulphur, possess sulphur-containing granules or globules while the pigments of photosynthetic bacteria are found under certain growth conditions as disc-shaped opaque or granular masses.

Staining techniques which demonstrate the existence of a nucleus in plant and animal cells have also been used to demonstrate nuclear material in bacteria. The presence of a true nuclear membrane, however, has not been demonstrated in bacteria. The nuclear material of bacteria is now known to exist as a continuous loop ramifying in the cytoplasm and indeed electron micrographs have been made showing the isolated bacterial chromosome, in the form of such a loop.

In addition some nuclear material with genetic properties can lie outside the main loop. Bacterial extrachromosomal elements exists in two forms, the first called episomes may at times occur in the main chromosomal loop and at other times extrachromosomally, the second called plasmids, never, as far as is known, ever become associated with the main chromosome.

Although plasmids may transfer any genetic property, one

property of great significance to medicine is the transference of resistant properties to antibiotics and other antimicrobial agents from one organism to another.

Such plasmids have been named R (resistance) factors and are found particularly amongst the Enterobacteriaceae (p. 121), but are also found in other bacteria, for example in the staphylococci where plasmid transfer occurs by transduction (p. 35).

The structure of the DNA molecule which forms the nucleus of all living cells and of some viruses and the extrachromosomal elements already referred to is shown on p. 110. It will be noted that it contains in its structure four bases, adenine, guanine, cytosine and thymine. These four bases in different combinations may direct, in the protein synthesizing system described on p. 103, the assembly of a sequence of amino acids to manufacture a unique protein which may be structural or an enzyme.

Further synthesis of enzymes may then confer upon the cell a recognisable ability i.e. the synthesis of a pigment or the property of carrying out a particular metabolic pathway.

This code for assembling amino acids is universal in *all* life and this fact and its implication has been described as the central dogma of biology.

Thus the chromosomes i.e. DNA strands of a cell can determine whether an elephant, a staphylococcus or a dandelion will be produced from available starting materials, the amino acids.

An assembly of base pairs to spell out a single complete property is called an operon.

Bacterial Flagella

Many bacteria are seen to be actively motile and possess thread-like appendages whose number and arrangement around the cell is used as a diagnostic feature.

The demonstration of flagella has been traditionally accomplished by a variety of special staining techniques which require careful practice in order to obtain satisfactory results. More recently, phase-contrast and electron microscopy have been applied to the study of flagella and have confirmed the earlier findings based on staining methods.

Flagella are arranged around the bacterial cell in characteristic patterns; monotrichate bacteria possess a single flagellum situated, in the case of rod-shaped organisms, at one end or pole of the rod; amphitrichate cells possess two flagella, one at either pole, and lophotrichate bacteria possess a cluster of flagella usually at one pole only but occasionally at both. Some cells are flagellated over their whole surface and are termed peritrichous. Flagella are mainly found amongst the rod-shaped organisms and the vibrios, although coccoid forms with flagella have been described.

The chemical structure of the flagella of many bacteria differs from that of the cell itself in at least sufficient detail as to be able to give rise to distinct antigens (page 27). The significance of the H and O antigens, as the flagellar and the cell or somatic antigens are known, will be dealt with in Volume 5. Each flagellum appears to be anchored within the cytoplasmic membrane to a small granular body and although argument to the contrary has been advanced, flagella are considered to be responsible for the motility of those bacteria which possess them.

By gentle agitation of flagellated bacteria with small glass beads, the flagella may be detached and have been found to consist of fibres of a single protein. Bacteria deflagellated by this method cease to be motile but the flagella will regrow. A study of the action of chemical substances on flagellar regrowth provides a useful research tool for the study of potential inhibitors of the synthesis of a single protein.

In addition to the flagella two other types of surface strand are present in Gram-negative bacteria, notably the Enterobactericeae (p. 121). The first are called pili or fimbriae and the second F-pili or sex strands. F-pili are usually larger than pili and have a bulbous end. Pili and sex strands are composed of protein but their diameter is less than that of flagella. The latter function in the primitive sexual reproductive processes of bacteria (p. 33) whereby parts of the nuclear material are transferred from one bacterium to another via the F-pili which are hollow. Pili are associated with the process of haemagglutination, i.e. red blood cell clumping, caused by some bacteria. Some six different structural types have been recognised in bacterial pili.

Layers Outside the Cell Wall

In many bacterial species there are coatings outside the rigid cell wall which may consist simply of an aggregation of slimy material or a cover of sufficiently definite structural status to warrant a special name, and called a capsule.

These layers do not appear necessary for the life of the cell and under certain conditions of growth may not be formed. Some authorities refer to all extramural material as the slime layer, while others refer to it as a capsule whether it is in the form of loose slime or a well-defined cover.

Chemically, extramural layers have been shown to consist of polysaccharides, polyuronic acids, simple sugars and amino sugars, and polypeptides. Recently, polymers related to the teichoic acids have been identified in the capsules of type 6 and 34 pneumococci (page 132).

The presence of bacterial capsules and slime have been demonstrated by a variety of techniques but as many of these involve quite drastic treatment, including the use of heat or dehydrating agents, it is not surprising that doubt has been cast on their validity in demonstrating accurately the true morphology of the capsule and slime layer; attempts to use the electron microscope have also been criticised, as the technique required to prepare the specimen for examination gives rise to distortion and artefacts.

A method which is simple and gives highly satisfactory results consists essentially of mixing the bacteria to be examined with a good quality Indian ink; the particles of carbon in water, which constitute Indian ink, fail to penetrate capsule or slime and the appearance of an encapsulated or slime-invested cell under the microscope is of a central zone, the cell itself, a clear zone, the slime layer or capsule, and a homogeneous extracellular area of Indian ink.

The role played by the extramural material in the economy of the organism is not immediately apparent but its presence has several consequences. By forming a protection against phagocytosis it may contribute to the virulence of pathogenic organisms. It may also provide a protection against antibacterial agents.

The capsular materials of many types of *Diplococcus pneumoniae* (the pneumococcus) are carbohydrates and differ in their nature in different strains of this organism, resulting in the recognition of more than 70 distinct types of pneumococci.

The capsule of *Bacillus anthracis*, the causal organism of anthrax, has been shown to consist of a polyglutamic acid with a molecular weight of the order of 50,000. This capsular substance, together with extracellular toxins, have been identified as the responsible agents for the pathogenicity of this organism.

The formation of capsules has been suggested as due to the accumulation of waste metabolic products and as such is a manifestation of metabolic inefficiency. *Aerobacter aerogenes* and the so-called mucoid strains of *Escherichia coli* produce a gummy carbohydrate, the amount produced being dependent on the cultural conditions. Thus, excess carbohydrate coupled with phosphate or nitrogenous deficiencies in a medium in which *A. aerogenes* is grown, increase the amount of the loose slime produced.

Leuconostoc dextranicum and *Leuconostoc mesenteroides* appear to have an inefficient mechanism for metabolising sucrose; these organisms first hydrolyse sucrose to glucose and fructose, but are unable to utilise glucose and polymerise it to form a complex carbohydrate known as dextran. This reaction represents an extreme example of extramural slime production and is exploited to produce a dextran for use in medicine as a substitute for plasma.

THE BACTERIAL SPORE

Certain bacterial genera, notably the rod-shaped *Bacillus* and *Clostridium*, are able to form structures within the vegetative cell called endospores or spores. These are far more resistant to heat and to other adverse environmental conditions than either the corresponding vegetative forms from which they arise, or non-spore-forming organisms. All sterilising procedures in pharmacy, medicine and the food industry have to take into account the resistant bacterial spore, when evaluating the efficiency of various processes.

The changes in microscopic appearance which accompany sporulation are an increase in refractivity or brightness. In addition, spores do not stain as readily as the corresponding vegetative form, but, once stained by heating with the stain, are much more difficult to de-stain.

The structure of the bacterial spore (Fig. 9) has been revealed in more detail with the advent of the electron microscope and the perfection of a technique for cutting sections of spores which may be examined in the electron microscope. Structurally the spore consists of an outer coat which is often constituted of concentric layers (up to four such layers have been seen in certain species, some species contain only two). The outer coat may be characteristically sculptured, a feature which has been strikingly demonstrated by depositing a layer of carbon on the spore and examining it with the electron microscope (Fig. 10). Concentrically, within the spore coat or coats, lies a structure called the cortex which may be

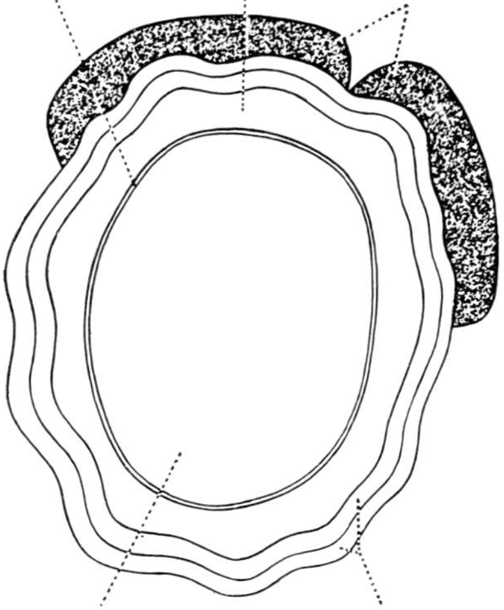

FIG. 9. Diagram of a transverse section of a bacterial spore

the barrier of permeability to aqueous solutions. Within the cortex lies the spore wall and within this wall the cytoplasm or spore core which contains the nuclear material of the vegetative cell from which it was formed. Spores of some species have a membrane external to the cell wall, and called the exosporium.

The size and position of the spore in relation to the vegetative cell from which it derived, that is whether it is larger equatorially than the parent vegetative rod, whether it is situated at the end (terminal), centre (central) of the rod, or between these two positions (sub-terminal) is often characteristic of a bacterial species and may be used as a diagnostic factor in bacterial classification (Fig. 11).

It has been stated that endospores are formed as a result of an adverse environment, for example, through lack of sufficient nutrient material or lack of moisture. This statement has been criticised as an oversimplification and more recent studies have suggested that the exhaustion of a specific substance or substances which may be organic or inorganic and which may vary from species to species, is responsible for initiating sporulation. Thus alanine limitation will initiate sporulation in some strains of

FIG. 10. Spores of *Bacillus polymyxa*. (Dr. D. E. Bradley, Electron Microscopy Laboratory, Dept. of Zoology, University of Edinburgh)

Bacillus mycoides and oxygen limitation will cause some strains of *B. megaterium* to spore. Other deficiencies that have been shown to induce sporulation include those of zinc, iron and phosphate ions. In addition, other constituents must be present for the spore to be formed. For example, many species require a source of calcium ions, while *B. subtilis* has been shown to require a source of manganese ions for spore formation. If the requisite substances for sporulation are not present, the cells in the culture

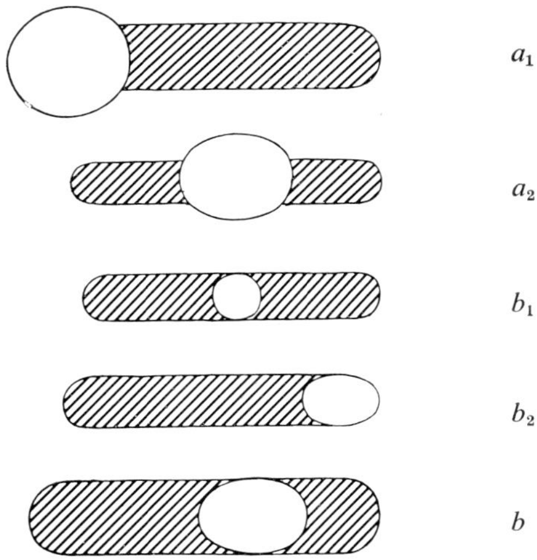

FIG. 11. Diagrams to show some variation of spore size and location with respect to the vegetative form from which they arise

a_1 Terminal spore larger than parent vegetative form
a_2 Central spore larger than parent vegetative form
b_1 Central spore of same size as parent vegetative form
b_2 Terminal spore of same size as parent vegetative form
b_3 Subterminal spore of same size as parent vegetative form

may die without spore formation occurring. Sporulation involves metabolic changes and the production of new chemical compounds which can be detected in the spore but not in the vegetative form; dipicolinic acid, pyridine 2,6-dicarboxylic acid is an example of such a compound (see below).

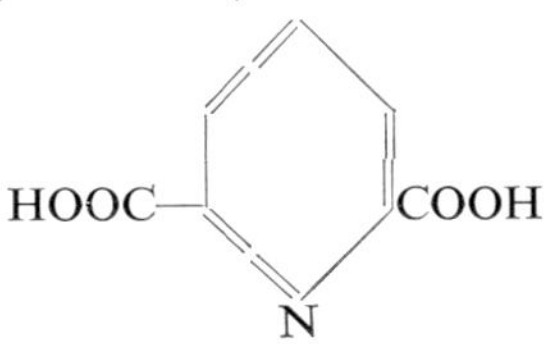

It is generally accepted that spores have a lower water content than vegetative cells. The low water content of the spore and especially of the core and the way in which the water present in the spore is bound have been cited as a reason for the heat resistance of this structure. In addition an interesting correlation has been shown to exist between heat resistance of spores and their dipicolinic acid and Ca^{++} content; heat resistance is associated with a high content of dipicolinic acid and calcium ions. It has also been suggested that the heat resistance of the spores lies in a special structural form assumed by spore protein during sporulation in which the protein molecules are cross-linked by calcium dipicolinate molecules to produce a heat-resisting structure.

Analyses of spores have shown that they have a higher cystine content than the corresponding vegetative form. Although some workers have sought to show that there is a significant correlation between cystine content and the radiation resistance of spores, this has also been disputed.

Much work has been carried out to resolve these theories but the spore nevertheless remains a resistant form of a micro-organism and eventually the full story of the resistance of the bacterial spore is emerging and appears not to be general for all spore-forming species.

The change from spore to vegetative form occurs in two stages, the first called germination and the second called outgrowth. (The term germination is often applied to the entire process.) During germination (the first stage) there is a loss of refractivity, an increase in ease of staining with simple stain procedures, and a loss of heat resistance together with a decrease in dry weight. Outgrowth is accompanied by increase in size of the spore, the splitting of the spore coat and emergence of the vegetative form, which may then if the external conditions are favourable begin to divide. The spore coat may split, equatorially or near the pole of the cell; this process is often characteristic of a given bacterial species. Spores may germinate, and subsequent outgrowth and multiplication occur when placed in a favourable environment, and it has also been demonstrated that germination will occur spontaneously when the spores are subject to a physical treatment or by the addition of a single chemical substance.

Chemical germination stimulants include L-alanine and glucose. Physical methods of inducing germination include heating spores to 60° C for one hour, the so-called sub-lethal heating, shaking with small (40 nm diameter) glass beads or treating with surface-active agents.

Subsequent outgrowth and development of the vegetative culture may require the usual array of nutrients dependent on the nutritional requirements of the organism concerned. Thus, in nature, the favourable environment implies the possible presence of a specific germination stimulator and adequate nutritional and environmental factors for vegetative growth.

In studies with the bacilli, it has been shown that the loss of weight of the spore on germination may amount to about one-third of the total dry weight of the spore. The weight loss has been shown to be due to the excretion from the spore of specific chemical substances which include calcium ions, dipicolinic acid and mucopeptides.

Spores of certain bacterial species have been shown to germinate at quite low temperatures, for example as low as 8°, and there is no insurance that keeping a product at a low temperature will prevent spores from germinating. Whether, once the spores have germinated and outgrowth has occurred, the vegetative culture formed from it will continue to flourish, is quite another matter; as has been said, conditions much more favourable than those which can initiate germination may have to be provided.

BACTERIAL PIGMENTS

The presence of photosynthetic pigments necessary for the life of one group of bacteria has already been mentioned. In addition many bacterial species produce pigments the function of which is not precisely known, but like the extramural layers already mentioned they may be evidence of metabolic inefficiency. Pigment production is used to a limited extent as a diagnostic feature in systematic bacteriology and some bacteria have been named according to the pigment they produce, for example *Staphylococcus aureus* (gold-coloured pigment), *Staphylococcus citreus* (lemon-yellow pigment).

The production of pigment by a bacterial species depends on cultural conditions; in general presence of oxygen, the use of solid culture media, rich in carbohydrate but with a low nitrogen content, an incubation temperature of 25°–30° C and growth in absence of light or in diffused light favour pigment production. Pigments may be classified into two groups, extracellular and intracellular, according to their occurrence outside or inside the cell. Some pigments are excreted into the surrounding medium. The main chemical groups to which the pigments belong are the carotenoids, quinones, phenazines and melanines. The red pigment of *Serratia marcescens*, the organism used to test bacteria-proof filters, is a tripyrrole derivative.

BACTERIAL LUMINOSITY

Several bacterial species are capable of emitting light. Many occur in the sea and are responsible for the luminosity of decaying fish on which they can proliferate.

Bacterial luminosity, as in the case of similar phenomena in fire-flies and glow-worms, is due to the presence of a substance called luciferin. In bacteria, luciferins have been shown to be long chain aldehydes such as decanal and dodecanal. This, in its reduced form, is oxidised by an enzyme luciferase. Luciferase, now reduced, is reoxidised by oxygen and during this process light is emitted. ATP (page 87) is also involved.

The first bacteria to evolve were anaerobes and as oxygen began to appear in the atmosphere some mechanism was necessary to destroy it if the anaerobes were to survive. Bacterial luminescence is thought to be a vestigial oxygen-reducing (removing) mechanism now playing no part in the economy of the cell.

Luminous bacteria have been used in an assay method for antibiotics, the end-point being taken at the point where the antibiotic caused a cessation of luminosity. The luciferin/luciferase reaction is also adaptable as an extremely sensitive method of determining ATP.

BACTERIAL TOXINS

Of the many thousands of bacterial species known only a relatively small number are pathogenic, that is disease producing, to man, animals and plants. It was upon this group however, until about 1920, that the bulk of scientific, clinical and field studies had been focused.

Disease caused by pathogenic bacteria is in most instances attributable to the production of complex poisons and tissue-destroying enzymes called toxins, although in certain cases the mere presence of the micro-organism can cause adverse symptoms.

Toxins may be divided conveniently into two classes: exotoxins, which are liberated from the organism into the tissues of the infected organism or host or into the culture medium of the organism being grown under laboratory conditions, and endotoxins, which are only liberated when the bacterial cell undergoes autolysis in the host or culture medium. In general exotoxins are protein in nature, while endotoxins may consist of one or more of three macromolecular components, proteins, polysaccharides or phospholipids. Toxic proteins are antigenic and the corresponding antibody, the antitoxin, is used as a therapeutic agent. The role of toxins in immunology and the production of immunological products is dealt with in Volume 5 of this series.

Enzymes

Hyaluronidase acts upon hyaluronic acid, the natural intracellular cement in animal tissue. It causes tissue breakdown and is responsible for the tissue-invasive activity of bacteria.

Preparations of hyaluronidase are used in medicine to promote the spread of medicament following their injection. Hyaluronidase is produced by staphylococci, streptococci and many clostridia.

Lecithinases hydrolyse lecithin and can cause haemolysis of red blood cells. Lecithinases are produced by *Cl. perfringens* (*welchii*) and *Cl. oedematiens.*

Proteinases hydrolyse protein. In the body they destroy muscle tissue and collagen. *Cl. perfringens* produces potent tissue-destroying enzymes.

Streptokinase occurs in certain streptococci; it can activate enzymes that in turn can dissolve blood clots. Its precise role in the pathogenicity of the organism is not clear.

Streptodornase, already referred to above, is produced by streptococci and can decompose deoxyribonucleoprotein and deoxyribonucleic acid (page 95). A preparation containing streptodornase is in use in medicine to liquefy blood and pus.

Coagulase, an enzyme produced by many staphylococci, increases the normal blood-clotting rate. The possession of coagulase activity is often taken as an index of pathogenicity in staphylococci but the exact role of coagulase in the pathogenicity of this organism is not at all clear.

Other Toxic Substances and Complexes

Cl. botulinum toxin. Cl. botulinum secretes a poison which is one of the most toxic substances known. It is produced extracellularly and death follows the ingestion of food in which this organism has been able to grow. The organism is a strict anaerobe and the type of environment in which it can flourish (with fatal results) is, for example, in canned meat in which the sterilisation process applied was inadequate or in meat products which are eaten without further cooking. The toxin is heat-labile, being destroyed by heating at 80°–100° C for 20 minutes.

Cl. tetani toxin. This anaerobe produces a potent nerve poison, and one of the clinical manifestations of infection is an inability to open the mouth—lock-jaw.

Cl. perfringens (*welchii*), *Cl. novyi* (*Cl. oedematiens*), *Cl. septicum toxins.* These three anaerobes are associated with the clinical condition known as gas gangrene. This group of organisms produces enzymes capable of destroying tissue, for example hyaluronidase and proteinases, and specific toxins with well-marked actions on differing tissues. *Cl. perfringens* toxins have been distinguished by prefixing Greek letters; examples are:

α-toxin—possesses lecithinase activity
θ-toxin—causes haemolysis and is a heart poison
κ-toxin—a proteinase
μ-toxin—hyaluronidase.

Toxins from B. anthracis. Recently it has been shown that a toxic complex of three components is produced by this organism. This toxin effects the permeability of blood capillaries causing internal haemorrhages sufficient to cause death. Animals dying of anthrax present the symptoms of secondary shock.

Toxins from staphylococci. Agents called leucocidins, capable of destroying white blood cells, are produced by many staphylococci. The production of a coagulase has already been mentioned but its role in the pathogenicity of this organism is doubtful.

Toxins from haemolytic streptococci. This group also produces toxins destroying white cells and red cells and an erythrogenic toxin which produces a characteristic red rash, as seen in cases of scarlet fever.

Diphtheria toxin. The causal organism of diphtheria, *Corynebacterium diphtheriae*, produces a toxin which interferes with protein synthesis. Heart muscle is particularly sensitive.

Cholera toxin. Causes an almost invariably fatal loss of fluid from the body tissues into the small intestine. This, in turn, causes a violent diarrhoea and dehydration of the body. Recently it has been shown that cholera toxin activates an enzyme called adenyl cyclase in the intestine which in turn induces water retention in the intestine. Its activation produces the results described.

Toxins from pathogenic Gram-negative bacteria. Both the composition of the toxins and the effects they produce are complex; amongst the organisms producing toxins are *Salmonella spp.*,

Shigella spp., *Neisseria gonorrhoeae* and *N. catarrhalis*. The toxins are comprised of a variety of macromolecules such as phospholipids, proteins and polysaccharides, and are released when the organisms undergo autolysis. That is they are endotoxins. The symptoms produced by infection by these organisms may include diarrhoea, haemorrhages, fever, changes in blood sugar levels and oedema, and to clinical entities such as gonorrhoea, bacillary dysentery and typhoid fever. Bacterial pyrogen is derived from Gram-negative organisms and consists of lipopolysaccharide.

BACTERIA AS ANTIGENS

When proteins, lipoproteins and some polysaccharides are injected into the blood stream of man or animals, specific substances are produced which react with or neutralise the injected substance. The injected substance is called an antigen, the substance produced is called an antibody. Some carbohydrates can act as antigens only in combination with a protein; such carbohydrates are known as haptens.

The carbohydrates of the capsule of *D. pneumoniae* act as haptens and distinguish the various pneumococcal types mentioned on page 17 and page 132.

Bacteria can act as antigens by virtue of their cellular constituents and this fact is exploited in the preparation of a variety of immunological products for use in medicine (Volume 4 of this series). It is also possible in some instances to determine if man or animals have been infected at any time with a particular micro-organism, by seeking antibodies to that organism in the blood. The phenomenon is also used for identifying micro-organisms and distinguishing closely allied species. To do this, an antibody to the particular organism must first be prepared. A suitable animal, usually the rabbit, is injected intravenously with the organism concerned which is usually killed by heat under such conditions that its activity as an antigen is not impaired. Time is allowed for the corresponding antibody to be formed. Blood is then withdrawn from the rabbit, allowed to coagulate and the serum, which contains the antibody, separated.

The characteristic reaction between antibody and a bacterial suspension containing the same antigen is agglutination or clumping of the cells in the suspension.

Agglutination may be observed by suspending the organism or

organisms grown on a solid medium in a drop of normal saline on a microscope slide or a glass plate and examining for absence of spontaneous agglutination. A drop of the agglutinating serum (the rabbit serum prepared as above) is now mixed with the bacterial suspension and the mixture examined. Agglutination is usually observable with the naked eye but may be confirmed with a hand lens.

Further confirmation of the identity of an unknown organism may be made by means of an agglutin in absorption test. Briefly the unknown organism is added in excess to the agglutinating anti-serum; during the agglutination process the antibody or antibodies will be absorbed from the serum. The mixture is now centrifuged and a sample of the serum from the clear supernatant layer is again tested against an authentic strain of the organism; agglutination should not now occur.

It has been found that micro-organisms which are indistinguishable biochemically and morphologically may be quite distinct antigenically. It has already been mentioned that the difference in structure of bacterial cells and flagella is sufficient to elicit the formation of different antibodies.

By the technique of serological typing not only may bacteria be classified and identified rapidly, but fundamental information concerning their chemical nature may be obtained.

Much use has been made of serological methods in distinguishing and identifying members of the Enterobacteriaceae and the streptococci using these extracted antigens (page 121).

THE STAINING OF BACTERIA

Bacteria are not intrinsically sufficiently coloured or refractive to be seen easily unless they are first coloured by a suitable dye, or unless special optical techniques are used.

In 1875 Weigert used the new synthetic dyes to stain bacteria and since then a variety of both dyes and ancillary reagents have been used in the study of bacteria, their appendages and structures.

It is not proposed to give protocols and recipes for bacterial staining here but to consider some factors of bacterial structure which affect the reaction of bacteria to stains.

Simple Staining Procedures

Bacteria may be stained by one of the following mechanisms.

1. Simple solution of the dye in a cell constituent, for example, a

lipid-soluble dye will dissolve in and stain lipid constituents of the cell.

2. Chemical reaction between dye and cell constituents. The residual amino or carboxylic acid groups of cell proteins may react with an acid or basic dye respectively. Triphenyltetrazolium chloride is reduced to a red triphenylformazan by reducing systems in a living cell and enables their position to be located.

3. Physical interaction based on Van der Waals attractions or on an ion-exchange mechanism may be responsible for some staining reactions.

Whereas mechanisms 1 and 2 are straightforward to comprehend, the mechanisms involved in the third category are less obvious. Two factors must be borne in mind, firstly that nearly all the dyes used in biology are either acids or bases, and secondly that under normal circumstances, the bacterial cell bears a negative charge, because there is an excess of groups, such as the —COOH group of proteins, in or near the cell surface. Thus the cation of a basic dye salt of the general structure R^+X^- (where R^+ represents the dye cation and X^- the corresponding anion, for example Cl^-) will react with the negatively charged surface of the cell and stain it. This process may continue as one of ion exchange, where the dye cation displaces cations from the cell surface.

Complex Staining Procedures

The Gram stain. There is one staining procedure, the Gram-staining method, which has probably received more attention both in mode of procedure and theoretical study than any biological staining technique in use today. It was first described by Christian Gram in 1884 and was evolved to detect bacteria in tissue and in dried preparations. Setting aside numerous modifications, the technique is as follows: the bacteria are spread on a microscope slide and fixed there by gentle heat. They are then stained in a solution of a basic dye such as crystal violet, followed by treatment with a solution of iodine. This is the actual staining technique. Bacteria can be divided into two classes by determining whether cells so treated may be subsequently decolorised by means of an organic solvent such as ethanol or acetone. Those organisms which are not destained are termed Gram-positive, those from which the stain may be washed are termed Gram-negative. After the destaining stage, the cells are treated with a second stain of a contrasting colour to the crystal violet such as a solution of fuschin, a red dye. Thus the final picture in the Gram-staining procedure is that Gram-positive cells appear blue and Gram-negative cells take the colour of the second stain or counter stain only, and in the example quoted appear red. At first sight it is strange that this series of operations should divide and classify bacteria into two categories, such a division being further paralleled by other well-marked differences.

Attempts to explain why in one group of bacteria, after treatment with a dye and iodine, the colour may be washed from the cell with ethanol and in another group it cannot, may be divided into two categories.

In the first category it is suggested that there is a special cellular component which after treatment with the dye and iodine so effectively combines with the dye that its subsequent solution in the decolorising solvent does not occur. Cellular components that have been suggested are lipids, nucleic acids, carbohydrates, lipo- and nucleo-proteins and glycerophosphates. In the second category, a physical theory, it is suggested that in bacteria from which the colour cannot be washed, the effect of the solvent is first to reduce the effective pore size of the cell wall and as a consequence of this the dye cannot pass out of the cell.

There have been many objections to the chemical theory in that, to put it at its simplest, too many exceptions to the rules can be quoted as a growing knowledge of cellular components accumulates. At the moment the most attractive hypothesis is that which cites changes in permeability induced by the decolorising solvent and the varying susceptibility of cells to this change.

Acid-fast organisms. A group of bacteria have been recognised since 1862 which are with difficulty stained with simple dye solutions such as fuschin but are so strongly stained if they are warmed with the dye that they resist subsequent destaining even by treatment with hydrochloric, nitric or sulphuric acid. Such bacteria are termed acid-fast and, because of the association of Ziehl and Neelsen with the development of the method, the staining regime to demonstrate the phenomenon of acid-fastness is called the Ziehl–Neelsen method. The method consists of heating a heat-fixed smear of the preparation to be examined with basic fuschin solution containing some phenol, until steam is seen to be rising from the slide. The slide is washed with 20 per cent sulphuric acid solution and then in water; it is then treated with a second contrasting dye such as methylene blue and re-examined.

Acid-fast organisms appear red, non-acid-fast organisms appear blue.

The classical example of an acid-fast organism is the causal organism of tuberculosis, *Mycobacterium tuberculosis*. Other members of the genus *Mycobacterium*, bacterial spores and many fungal spores are acid-fast and a similar technique can be used to differentiate spores from vegetative forms in a bacterial culture that might contain both.

Acid-fastness in the mycobacteria has been attributed to the presence in the cells of waxes which strongly combine with the stain when heated with it. Later a permeability theory was put forward, as it was found that only intact cells were acid-fast and that the waxes extracted from mycobacteria were not. The heating process was thought to allow the dye to enter the cell and subsequently seal it in so that it was not re-extractable by acids.

BACTERIAL REPRODUCTION

An outline of the cytological events which accompany cell division will be given in this section. The pattern of reproduction of populations of bacterial cells is given in Chapter 2.

The non-filamentous or non-branching bacteria divide by simple binary fission, that is one cell divides to give two cells and these two each divide again to give four cells and so on. The generation time, that is the time for one cell to divide to give two cells, is in many species of the order of 20 minutes. The nuclear material replicates itself and the two replicas segregate at opposite sites in the cell and this is followed by the formation of a new cell wall. In Gram-negative rod-shaped organisms an annular constriction first appears, approximately midway between the ends of the cell which, by a progressive decrease in circumference accompanied by the formation of new wall material, finally gives rise to two cells. The cells may remain adherent, giving on further division chains of cells, or they may separate.

In Gram-positive cocci, after nuclear segregation, equatorial thickenings or ridges of the cell wall appear on the interior of the wall. These ridges now grow across the equatorial plane of the cell like a pair of closing iris diaphragms until finally two separate hemispherical cells are formed. The two cells thus formed may separate and become individual spherical cells but separation does not always occur. Failure to separate gives rise to the diplococcal, streptococcal, staphylococcal and sarcinal forms already mentioned. The shape of these aggregates is determined by the relative positions of the planes of the equatorial cross walls of the second and subsequent divisions. Gram-positive rods divide by forming cross-walls rather than by the process of constriction seen with Gram-negative rods. Reproduction of the bacterial genus *Streptomyces* is by means of spore-like structures or conidia (page 140).

It might well be inferred from the above description that reproduction amongst the bacteria was a strictly asexual process and that a bacterial species was immutable or, put in another way, the phenomenom of genetic recombination, which is an invariable possibility in the strictly sexual reproduction of higher organisms, cannot occur.

Nevertheless the phenomenon of bacterial variation referred to in the bacteriological literature from the 1880's onwards suggested that certain characters of bacterial species could change, and some of these changes were inheritable. Examples of such changes include changes in cell morphology and often as a result a change in the appearance of bacterial colonies. Thus a change or variation involving loss of a capsule may result in a characteristic change in colony appearance from a smooth to a rough colony, the so-called S → R variation. An excess of extramural slime may be produced in an organism which normally does not produce such material, giving rise to mucoid colonies or a mucoid variant. Ability to produce pigment is another variable quality. Another type of cellular variation is a loss of flagella and hence motility. Variation in virulence and in antigenic structure is also found. Variation in nutritional requirements is often encountered.

While it is undoubtedly true that some changes in the characters of micro-organisms are due to adaptive changes which may not be genetically controlled, it is now known that change and exchange in the genetic make-up of bacteria can and does occur.

Changes in the genetic make-up of cells can be brought about by ultraviolet light of certain wave-lengths and this might be one of the processes of natural or spontaneous mutation. If the mutant is more suited to its environment than the original organism it may, by a process of natural selection, become the predominant and finally the only strain in the environment.

Conversely the mutation might be lethal, depriving the organism of an essential metabolic activity, thus resulting in death.

Treatment by X-rays and mutagenic chemicals such as nitrogen mustard (di-(2-chloroethyl)methylamine) has been used to produce artificially new mutants of both bacteria and moulds and to obtain strains of these organisms with desired properties. Mutants of the mould *Penicillium chrysogenum* have been produced which give much higher yields of penicillin and at the same time do not produce the unwanted yellow pigment associated with the original strain.

Genetic Exchange in Bacteria

The fundamental characteristic of sexual reproduction is that it allows the exchange of genetic material between the mating pair. In 1946 an exchange was demonstrated in bacteria by Lederberg and Tatum and since this experiment three patterns of genetic exchange have been recognised, confugation, transformation and transduction.

Bacterial conjugation (the 1946 experiment of Lederberg and Tatum). *Escherichia coli* will grow on a simple medium of inorganic salt and glucose (Chapter 2) and such are its biosynthetic capabilities that it can synthesise all the requirements for proteins, nucleic acids and fats from this minimal medium. Treatment with mutagenic reagents or X- or ultraviolet rays can produce strains or variants which have lost their ability to synthesise one or more essential cellular components, for example one or more amino acids, and, unless these are now supplied in the minimal medium, the newly created strain cannot grow. It is assumed that each nutritional, or for that matter any other, deficiency is due to the loss of function of a single gene, caused in this case by the mutagenic treatment. The essence of the Lederberg and Tatum experiment was firstly the obtaining of a strain of *Esch. coli* requiring factors which may be called *b*, *c* and *d*, and secondly a strain requiring *e* and *f*; thus the first strain differs from the parent non-exacting strain in lacking the ability, through gene deficiency, of synthesising *b*, *c* and *d* and the same argument applies to the second strain with respect for *e* and *f*. The two deficient strains were grown together in a medium supplied with the factors *b*, *c*, *d*, *e* and *f*; thus all grew. After growth, some of the cells were transferred to the minimal medium, when about one in each million cells transferred could grow. It was thus inferred that in some way material containing *b*, *c* and *d* genes had been transferred from an *e*, *f* cell to a *b*, *c*, *d* deficient cell or *e* and *f* genes had been transferred to an *e*, *f* deficient cell.

The mechanism of this exchange process has now been elucidated in some detail.

Firstly it was found that not all strains of *Esch. coli* would participate in this exchange process and Hayes showed that it occurred only in those organisms containing in their genetic make a fertility factor, F. Strains possessing such a factor were designated F^+ and strains without it, F^-. A genetic exchange can occur between F^+ and F^- strains. As further studies proceeded, a particularly efficient transferring strain of F^+ organisms was recognised and this was termed Hfr, the letters standing for high frequency recombination. Yet a third type of F strain has also been characterised. Written F' and read as F-prime.

These three fertile donor types, called by analogy with higher organisms males, differ in the extent and efficiency with which they transfer genetic material to F^- or 'female' recipients.

Possession of the fertility factor enables organisms to produce the sex hairs or sex pili already referred to and it is through these that genetic material is passed, hence their alternate name conjugation tube, but where F^+, F' and Hfr differ is in the location of the fertility gene in relation to the cells main genetic material, its annular chromosome.

In F^+ strains the F factor is present as an extrachromosomal element, i.e. separately from the chromosome. In F' strains the F factor is extrachromosomal but contains some genetic material derived from the main chromosome, but in Hfr strains the F factor is fully integrated into the chromosome.

Conjugation between F^+ and F^- results in the formation of F^+ cells. $F^+ \times F^-$ results in F^+ cells with some of the genetic make up of the donor while Hfr $\times$ F^+ can give rise to a complete transfer of all the genetic material originally present in the Hfr chromosome including F itself. This statement, however, needs qualifying.

The primitive mating process described above involved first the fusion of the sex hair of an F^+ strain with the surface of the F^- strain, followed by the transfer of nuclear material via its hollow interior. In the case of an Hfr/F^- mating, involving as it does, the whole chromosome the process may take up to $1\frac{1}{2}$ hr. Fracture of the sex hair may occur, even the energy inherent in Brownian movement may be sufficient, so that only part of the chromosomes enters the recipient cell. However, interrupted conjugation as it is called is not usually fatal and the progeny of the recipient may show in its properties partially expression of the genetic make-up of the donor. As a point of interest the annular chromosome of the Hfr donor fractures at the point where the F factor lies and this is the last factor of the now linear chromosome to pass or attempt to pass through the conjugation tube to the recipient.

Donor tube fracture may be indiced by mechanical means, i.e. by agitating Hfr and F′ cultures with glass beads. By taking samples of such a mixture and varying times during the $1\frac{1}{2}$ hr or so required for total conjugation, allowing them to grow and examining their properties, some notion of the order or location of genes along the chromosome responsible for a particular property may be deduced.

It can also be seen that in Hfr/F′ conjugation it is the F factor, the last, which has the greatest chance of not being transferred, the chance increasing the nearer the particular factors are to the head of the opened chromosome.

This is a very simplified picture of an important fact of bacterial genetics but serves to illustrate in outline this primitive pattern of sexual reproduction which contain the shadows of that to be evolved in the most advanced inhabitants of this planet.

Transformation. A second example of the transfer of genetic material has been found in *Diplocossus pneumoniae* (page 132), *Neisseria meningitidis* (page 131) and *Haemophilus influenzae* (page 129). Here it was found that inheritable characters and by inference genetic material may be transferred not by mixing culture as in the previous example but by adding a deficient culture to the filtrate from a culture in which cells not containing the deficiencies have been grown. This phenomenon is known as transformation, and was discovered by Griffith in 1928. Avery and his colleagues showed in 1944 that the transferred material was indeed DNA see p. 110.

Although Griffiths original discovery related to the transformation of capsule material in the pneumococcus it is now known that other properties including drug resistance may be transformed.

In this process no conjugation system is present and it is thought that the hereditary material escapes from the one set of cells possibly if they autolyse and may find their way into the recipient cells through the activity of autolytic enzymes which dissolve enough of the cell wall to allow the large DNA molecule to enter an otherwise impermeable cell.

Transduction. The third method called transduction was discovered in 1952 by Zinder and Lederberg. These workers found that a bacteriophage (bacterial virus) p. 149, could transfer bacterial genetic material from one organism to another.

The viruses capable of effecting this genetic transfer belong to a group known as temperate the feature of which is they do not harm their host bacterium.

This process of genetic transfer is called transduction but it should be remembered that only a very small amount of the genetic bacterial material is transferred by this process.

References

The Bacteria. Vol. 1: *Structure.* Ed. Gunsalus, I. C., and Stanier, R. V. Academic Press, 1960

Bacterial Anatomy. 6th Symposium, Soc. gen. Microbiol. Ed. Spooner, E. T. C., and Stocker, B. A. D. Cambridge University Press, 1956

The Surface Structure of Bacteria. Rogers, H. J. In Biochemical Society Symposium No. 22. The structure and function of the membranes and surfaces of cells, p. 55. Ed. Bell, D. J., and Grant, J. K. Cambridge University Press, 1963

Spore Formation and Germination as a Microbial Reaction to the Environment. Murrel, W. G. In *Microbial Reaction to Environment.* 11th Symposium, Soc. gen. Microbiol., p. 100. Ed. Meynell, G. G., and Gooder, H. Cambridge University Press, 1961

Microbial and Molecular Genetics. Fincham, J. R. S. English Universities Press Ltd, 1965

Sexuality and Genetics of Bacteria. Jacob, F. and Wollman, E. L. Academic Press, 1961.

Cell Walls and Membranes. Rogers H. J. and Perkins, H.R. E. and F. Spon, 1968

The Bacterial Spore. Ed. Gould, G. W. and Hurst, A. Academic Press, 1969

CHAPTER 2

BACTERIAL GROWTH AND FACTORS AFFECTING BACTERIAL GROWTH

THE CULTIVATION OF BACTERIA

The bacterial cell has so far been considered as a single entity; normally, cells are seen collectively as aggregations of many cells on a solid medium which are known as bacterial colonies or in liquid culture conferring turbidity on an originally optically clear medium. The morphology of bacterial colonies is of importance in systematic bacteriology.

Bacteria grow in natural habitats such as the soil, water, the alimentary tracts of animals and on foodstuffs such as meat and milk, to quote only a few examples. Their presence in the soil is of practical significance. In solutions to be used parenterally, in preserved food, milk and public water supplies, the presence of bacteria is potentially dangerous and the purification of water and preservation of foods and pharmaceutical products is concerned with the elimination or killing of bacteria and other micro-organisms.

The chance cultivation of micro-organisms had occurred long before their nature or existence was proved; their growth in carbohydrate-containing media, for example grape juice, to produce alcoholic beverages and their growth on animal carcases producing putrefaction have been known from very early in man's history. For practical study in the laboratory, however, an array of culture media have been evolved, some designed to grow all types of bacteria, others to select from a mixed environment a particular bacterial species.

Most culture media used routinely are based on products made from meat extracts, or digests.

Partially evaporated aqueous extracts of ox muscle ('Lemco') or dried enzymic or acid digests of muscle tissue, casein or blood fibrin (peptones) form the basis of most culture media. The addition of agar (2 per cent) results in a solid medium, on which isolated colonies can be grown by spreading a culture thinly over the surface of the medium. A typical nutrient broth formula is meat extract (10 g), peptone (10 g), and sodium chloride (5 g) in water to 1 litre.

This may be converted to a solid medium by the inclusion of agar (20 g). A medium more suited to the growth of moulds may be prepared by including glucose or malt extract in the above recipe. The nutrient value of these media may be conveniently increased if necessary by adding a dried, or partly dried, hydrolysed or autolysed extract of baker's yeast to the extent of 0·1 per cent. Blood or serum is also used as an adjunct to some media.

There is for each bacterial species, a set of interrelated conditions which give optimum growth. These conditions are:

1. The supply and balance of solid nutrients.
2. The supply of gaseous nutrients, for example of oxygen, carbon dioxide or nitrogen.
3. Hydrogen ion concentration.
4. Temperature.
5. Osmotic pressure.
6. The presence of water.

BACTERIAL GROWTH

The Measurement of Bacterial Growth

The quantitative effect of the many variables mentioned above and to be considered below can be assessed only when the extent of bacterial growth can be measured. There are four main methods of measuring bacterial growth:

1. Total counts by direct counting of the bacteria when viewed under the microscope.
2. Viable counts, a technique which seeks to measure the living population of a bacterial culture by adding samples to culture media and counting the colonies assumed to be produced by the growth of each original cell in the sample.
3. Indirect counts which, using a suitable optical instrument, measure the turbidity or light-scattering properties of a suspension.
4. Biochemical methods which measure the increase of, for example, protein in the suspension, again assuming that this is increasing with increasing bacterial numbers. It should be clearly realised that methods 1, 3 and 4 measure either directly or indirectly the total bacterial population, whether living or dead.

The Pattern of Bacterial Multiplication

Assuming optimum or near optimum conditions, the pattern of the multiplication of bacteria from an inoculum into a liquid medium, is as follows (Fig. 12).

1. *An initial stationary or lag phase.* In typical instances this may be as little as 20 minutes, during which there is very little, if any, increase in the numbers of the viable population. The cells, however, are metabolically active in that they consume nutrients and increase in size, and are more susceptible to adverse conditions

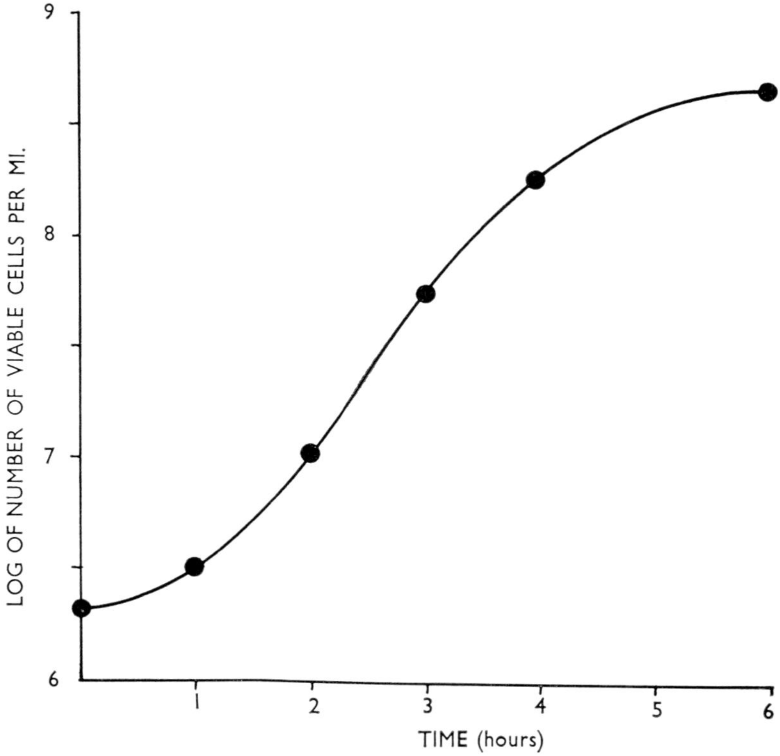

FIG. 12. Bacterial growth curve. *Escherichia coli* in nutrient broth at 37° C. (A. D. Russell, Ph.D. thesis, University of Nottingham)

such as extremes of temperature or the action of toxic substances than when actively growing.

2. *A log phase.* Following the initial stage, in which bacterial numbers are not increasing, there occurs a phase of cell division in which one cell gives rise to two, these two to two more, or put in general terms the bacterial numbers increase by geometrical progression. Thus, at this stage, there is a linear relationship between the logarithm of the number of viable bacteria and time. It should

be realised that in this stage every cell is not dividing, as it is also not strictly true to say that, in the stationary phase, no cell divides. Nevertheless, the general pattern of a lag and log phase is characteristic.

3. *A stationary phase.* Following upon the phase of active growth, there follows a phase when again little or no growth occurs; this is referred to as the stationary phase. The most obvious explanation of this situation is that it comes about because of exhaustion of essential food materials by the multiplying cells; a variety of experiments has suggested that other factors may be involved, such as the achievement of a toxic pH or the accumulation of a toxic metabolite. Some workers have suggested that for each culture there is a concentration level, the so-called *M*-concentration, at which a given bacterial species under identical conditions in a fluid culture attains the same maximum population, or alternatively the point where the lag phase ends and the stationary phase begins is a physiological constant.

4. *Phase of decline.* If viable counts are continued there is found to be a falling off of viable numbers. This period may be lengthy; viable organisms have been recovered from cultures of certain species, including those that do not form spores, in nutrient broth after periods at least as long as twelve months. This phase is not shown in Fig. 12.

Synchronous Growth in Bacterial Cultures

It has been discovered within the last ten years that bacterial cultures may be made to give a series of growth curves much more precise than those obtained under normal circumstances. Such a culture shows a phase of logarithmic growth, and a stationary phase, followed by a second phase of logarithmic growth. During the logarithmic growth phase, nearly every cell is thought to be undergoing division and such a culture is spoken of as a synchronously dividing culture. Use has been made of this technique to study the action of substances on a growing population of bacteria all considered to be of the same age.

One method of achieving these stepwise bursts of growth is to cool a growing culture from 37° to 4° C and then rewarm to 37°. The growth curves shown in Fig. 13 were obtained by such a method.

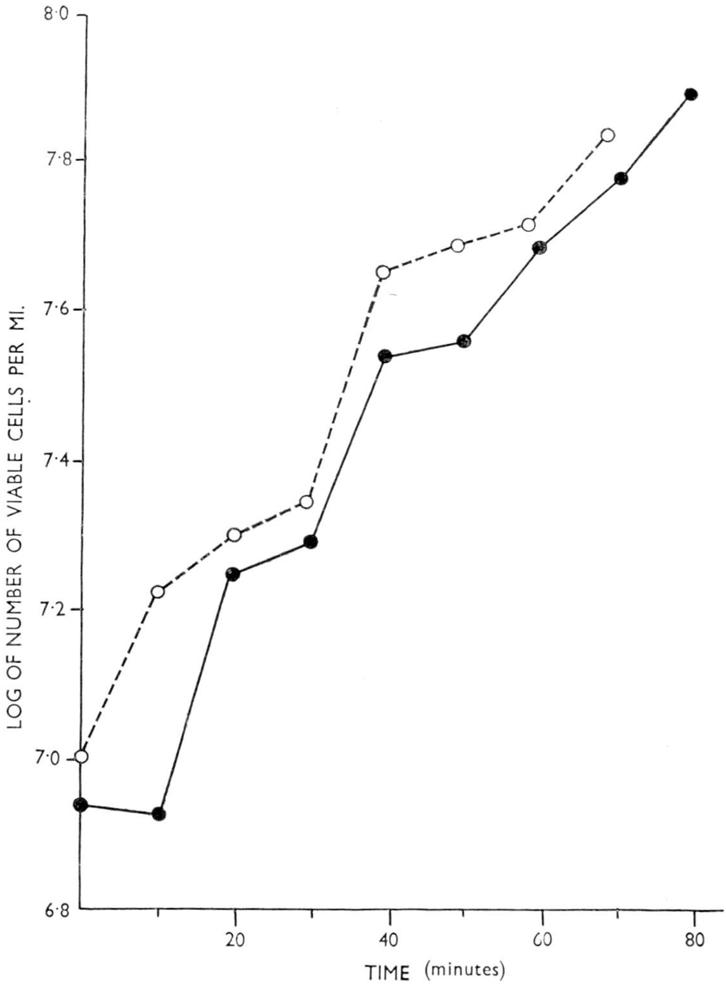

FIG. 13. Synchronous growth in a nutrient broth culture of *Escherichia coli*, Results of two separate experiments are shown. (A. D. Russell, Ph.D. thesis. University of Nottingham)

The above data have been obtained from what is called batch culture. In processes involving the production of a chemical substance employing micro-organisms, for example in autibiotic or vinegar manufacture, a process called continuous culture may be more economical.

Continuous Culture of Micro-organisms

In this process, spent culture is continuously removed and fresh media is continually added to produce, ideally, a steady state or continuous logarithmic state of growth, i.e. the stationary phase and phase of decline as seen in batch culture and discussed above is never seen.

Many ingenious pieces of automated equipment have been described to achieve this end.

Continuous culture has been the subject of much study and mathematical analysis. One mathematical parameter used in the dilution rate D which has the dimensions of a reciprocal hour, (h^{-1}).

$$D = \frac{\text{rate of flow of added nutrient (litres/hr)}}{\text{volume of culture vessel (litres)}}$$

$$= \frac{f}{v}$$

If flow rates are made to maintain a constant turbidity in the system the apparatus is often spoken of as a turbidostat, if however the concentration of a chemical in the medium, for example a substrate or reaction product, then the term chemostat is used.

Mean Generation Time

The time interval between one cell division and the next is called the generation time. When considering a growing culture containing many thousands of cells, a mean generation time is usually calculated.

If a single cell reproduces by binary fission, then the number of bacteria, *n*, in any generation will be as follows:

$$\begin{aligned} &\text{1st generation } n = 1 \times 2 &&= 2^1 \\ &\text{2nd generation } n = 1 \times 2 \times 2 &&= 2^2 \\ &\text{3rd generation } n = 1 \times 2 \times 2 \times 2 &&= 2^3 \\ &Y\text{th generation } n = 1 \times 2^Y &&= 2^Y \end{aligned}$$

For an initial inoculum of n_0 cells, as distinct from one cell, at the *Y*th generation, the cell population will be:

$$n = n_0 \times 2^Y$$

This equation may be rewritten thus:

$$\log n = \log n_0 + Y \log 2$$

whence

$$Y = \frac{\log n - \log n_0}{\log 2} = \frac{\log n - \log n_0}{0{\cdot}3010}$$

where Y = the number of generations that have elapsed in the time interval between determining the viable count n_0 and n.

If this time interval is t, then the mean generation time G is given by the expression:

$$G = \frac{t}{Y} = \frac{t}{\dfrac{\log n - \log n_0}{0{\cdot}3010}} = \frac{t \times 0{\cdot}3010}{\log n - \log n_0}$$

Instead of determining the mean generation time, a factor called the growth constant, k, may be determined.

During the logarithmic phase of growth the rate of increase of bacterial numbers, n, with time, t, is given by the expression:

$$\frac{dn}{dt} = kn$$

If the bacterial population has increased from n_0 to n in the time, t, then on integration the above expression may be written:

$$\log_e \frac{n}{n_0} = kt$$

This may be re-expressed and in terms of $\log_{10}$, thus:

$$\frac{(\log n - \log n_0)}{t} = \frac{k}{2{\cdot}303}$$

The factor 2·303 is introduced when logarithms to the base e are converted to logarithms to the base 10.

Thus $$k = \frac{(\log n - \log n_0) \times 2{\cdot}303}{t}$$

THE EFFECT OF ENVIRONMENTAL FACTORS ON THE GROWTH OF BACTERIA

Nutritional Requirements

Bacteria can utilise and may have an absolute requirement for the following substances:

Carbon. This may be in the form of an organic compound, such as glucose, starch and even phenol, or as a bicarbonate or carbon dioxide. All bacteria appear to need a small amount of carbon dioxide for growth; some can derive all their carbon requirements from a bicarbonate or carbon dioxide.

Hydrogen. This is usually derived from the breakdown of water. In a few cases hydrogen, derived from the breakdown of carbohydrates, may be remetabolised.

Oxygen. This is derived from the atmosphere. In the anaerobic bacteria free oxygen is not, of course, required.

Nitrogen. This may be utilised as elementary nitrogen, inorganic nitrogen, amino acids or amino acid polymers. Many bacteria can utilise, or as it is often termed 'fix', elementary nitrogen. This process is of great importance in that the nitrogen of the atmosphere can be made available for the nutrition of higher plants. Inorganic nitrogen as ammonium salts, nitrates and in certain cases nitrites can also serve as nitrogen sources.

Most bacteria can utilise amino acids (Table 1) as a source of nitrogen; some have an absolute requirement for one particular amino acid; others may need all their amino acids pre-supplied (Table 2). Amino acid polymers, i.e. proteins and peptides, are degraded by many micro-organisms to produce utilisable nitrogen-containing units.

Sulphur. A few species can use elementary sulphur. Inorganic sulphur such as hydrogen sulphide or sulphates are decomposed to render sulphur available to the cell in some species. The sulphur-containing amino acids, cysteine, cystine and methionine (organic sulphur), can serve as sulphur sources for many bacteria.

Phosphorus. As far as is known, inorganic phosphorus, usually as phosphates, serves as the phosphorus source in nutrition.

Growth factors. Examples are given on page 47. The use of the word 'growth factor' enables a miscellaneous group of chemicals to be classified functionally. Some texts would describe an essential amino acid as a growth factor. In this account it is proposed to use the term 'growth factor' in the restricted sense. Many of the substances given are the vitamins of human nutrition. It is of interest to note that if maximum growth is obtained at levels of 10^{-4} to $10^{-5}M$ the growth factor or amino acid is probably being built into the cell, whereas if the level for maximum growth is 10^{-8} to $10^{-10}M$ the substance concerned is acting as a catalyst, probably as part of an enzyme system.

Trace elements. As far as is known, all living organisms need traces of the salts of certain chemical elements and also certain

TABLE 1

COMMON AMINO ACIDS

Name	Abbreviation	Formula
Alanine	Ala	$CH_3{\cdot}CH(NH_2){\cdot}COOH$
Arginine	Arg	$H_2N{\cdot}C(:NH){\cdot}NH{\cdot}CH_2{\cdot}CH_2{\cdot}CH_2{\cdot}CH(NH_2){\cdot}COOH$
Aspartic Acid	Asp	$HOOC{\cdot}CH_2{\cdot}CH(NH_2){\cdot}COOH$
Cysteine	CySH	$HS{\cdot}CH_2{\cdot}CH(NH_2){\cdot}COOH$
Cystine	CyS	$S{\cdot}CH_2{\cdot}CH(NH_2){\cdot}COOH$ | $S{\cdot}CH_2{\cdot}CH(NH_2){\cdot}COOH$
Diaminopimelic acid	DAP	$CH{\cdot}NH_2{\cdot}COOH$ | $(CH_2)_3$ | $CH{\cdot}NH_2{\cdot}COOH$
Glutamic acid	Glu	$HOOC{\cdot}CH_2{\cdot}CH_2{\cdot}CH(NH_2){\cdot}COOH$
Glutamine	Glu(NH_2)	$NH_2.COCH_2.CH_2CH(NH_2).COOH$
Glycine	Gly	$CH_2(NH_2){\cdot}COOH$
Histidine	His	 $CH_2{\cdot}CH(NH_2){\cdot}COOH$
Isoleucine	Ileu	$CH_3{\cdot}CH_2{\cdot}CH(CH_3){\cdot}CH(NH_2){\cdot}COOH$
Leucine	Leu	$(CH_3)_2CH{\cdot}CH_2{\cdot}CH(NH_2){\cdot}COOH$
Lysine	Lys	$H_2N{\cdot}CH_2{\cdot}CH_2{\cdot}CH_2{\cdot}CH_2{\cdot}CH(NH_2){\cdot}COOH$
Methionine	Met	$CH_3{\cdot}S{\cdot}CH_2{\cdot}CH_2{\cdot}CH(NH_2){\cdot}COOH$
Phenylalanine	Phe	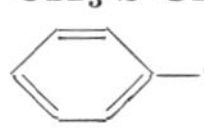 —$CH_2{\cdot}CH(NH_2){\cdot}COOH$
Proline	Pro	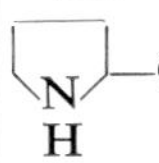 —COOH
Serine	Ser	$HO{\cdot}CH_2{\cdot}CH(NH_2){\cdot}COOH$
Tryptophan	Try	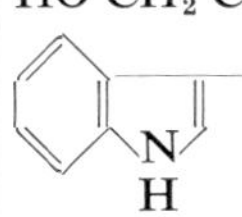 $CH_2{\cdot}CH(NH_2){\cdot}COOH$
Tyrosine	Tyr	HO— 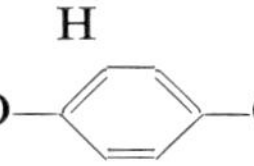 —$CH_2{\cdot}CH(NH_2){\cdot}COOH$
Valine	Val	$(CH_3)_2CH{\cdot}CH(NH_2){\cdot}COOH$

TABLE 2

NUTRITIONAL REQUIREMENTS OF SOME HETEROTROPHIC BACTERIA

	Esch. coli	*Sal. typhosa*	*Pr. vulgaris*	*Pr. morganii*	*N. meningitidis*	*Staph. aureus*	*Strep. faecalis*	*Strep. haemolyticus*	*Lb. casei*	*Lb. arabinosus*	*D. pneumoniae*	*C. diphtheriae*	*B. anthracis*	*Cl. sporogenes*	*Cl. acetobutylicum*	*Cl. tetani*
Glycine	—	—	—	—	—	+	?	?	?	?	+	?	+	—	—	—
Alanine	—	—	—	—	—	*	+	+	+	?	+	?	+	—	—	—
Serine	—	—	—	—	—	?	?	+	+	?	?	?	?	—	—	*
Cystine	—	—	—	+	+	+	?	+	+	?	+	+	+	?	—	—
Phenylalanine	—	—	—	—	—	+	?	+	+	?	+	?	+	+	—	*
Tyrosine	—	—	—	—	—	*	+	+	+	?	?	+	+	+	—	+
Tryptophan	—	+	—	—	—	?	+	+	+	+	+	?	+	+	—	+
Threonine	—	—	—	—	—	?	+	+	+	?	?	?	?	—	—	*
Valine	—	—	—	—	—	+	+	+	+	+	+	+	+	+	—	+
Leucine	—	—	—	—	—	+	+	+	+	+	+	?	+	+	—	+
Isoleucine	—	—	—	—	—	?	+	+	+	+	?	?	+	—	—	+
Glutamic acid	—	—	—	—	+	+	+	+	+	+	+	+	+	—	—	—
Aspartic acid	—	—	—	—	—	+	+	+	+	?	+	+	+	—	—	—
Histidine	—	—	—	—	—	+	+	+	+	?	+	?	+	?	—	+
Lysine	—	—	—	—	—	*	+	+	+	?	+	?	+	—	—	—
Arginine	—	—	—	—	—	+	+	+	+	?	+	?	+	+	—	+
Methionine	—	—	—	+	—	*	+	+	+	+	+	+	+	+	—	—
Proline	—	—	—	—	—	+	?	+	?	?	?	+	+	—	—	—
Hydroxyproline	—	—	—	—	—	*	?	?	?	?	+	?	+	—	—	—
Glutamine	—	—	—	—	—	?	?	+	+	*	+	+	?	?	?	?
Asparagine	—	—	—	—	—	?	?	+	+	?	+	+	—	—	+	+
Nicotinic acid	—	—	+	+	—	+	?	+	+	+	+	+	—	?	—	+
Thiamine	—	—	—	—	—	+	—	?	?	?	?	?	—	?	—	+
Riboflavin	—	—	—	—	—	—	+	+	+	+	+	?	—	?	—	+
Pantothenic acid	—	—	—	+	—	—	+	+	+	+	+	+	—	?	—	+
Pyridoxal	—	—	—	—	—	—	+	+	+	—	?	?	—	?	—	+
Biotin	—	—	—	—	—	*	*	+	+	+	+	?	—	+	+	+
Pimelic acid	—	—	—	—	—	?	?	?	?	?	?	+	—	?	?	?
p-Aminobenzoic acid	—	—	—	—	—	—	—	*	+	+	?	—	—	?	+	—
Pteroic acid	—	—	—	—	—	—	+	?	—	—	?	—	—	—	—	—
Pteroylglutamic acid	—	—	—	—	—	—	—	+	+	*	*	—	—	?	—	+
Adenine	—	—	—	—	—	—	+	+	+	+	+	?	—	?	—	+
Uracil	—	—	—	—	—	+	—	+	?	+	?	—	—	?	—	+
Purines } Pyrimidines }	—	—	—	—	—	—	+	+	+	+	+	?	—	+	—	+

KEY: + essential for most strains
* essential for some strains
— not essential
? unequivocal data not available

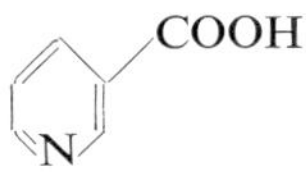

Nicotinic acid

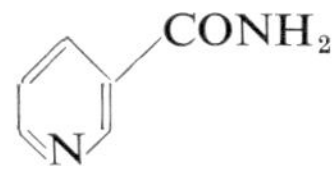

Nicotinamide

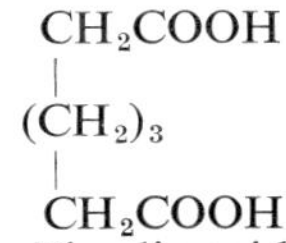

Pimelic acid

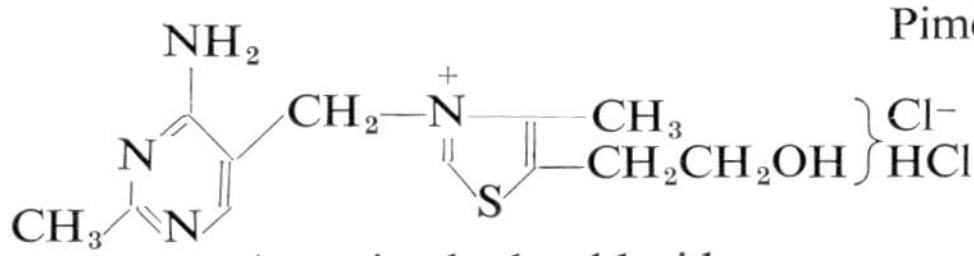

Aneurine hydrochloride
(Thiamine hydrochloride; Vitamin B_1)

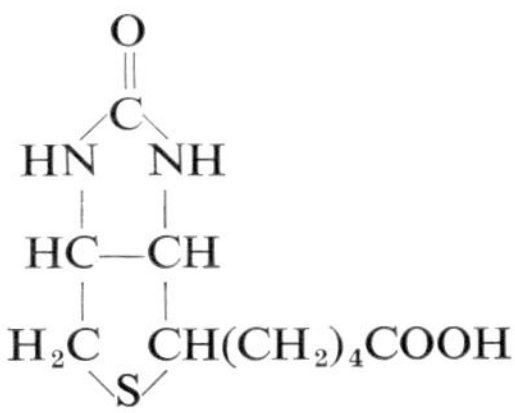

Biotin

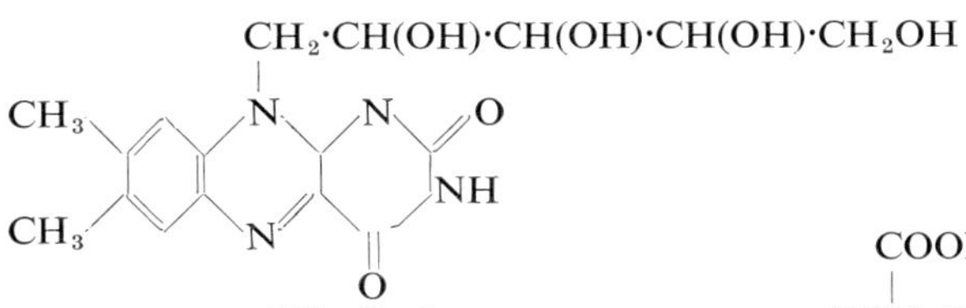

Riboflavine
(Vitamin B_2)

Pantothenic acid

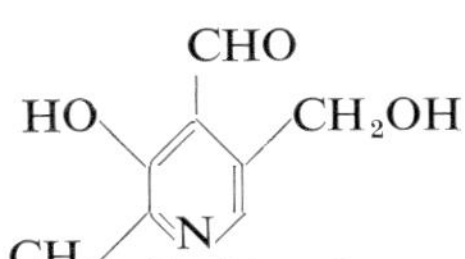

Pyridoxal
(Vitamin B_6)

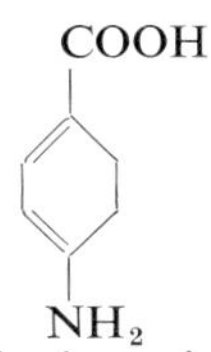

p-Aminobenzoic acid

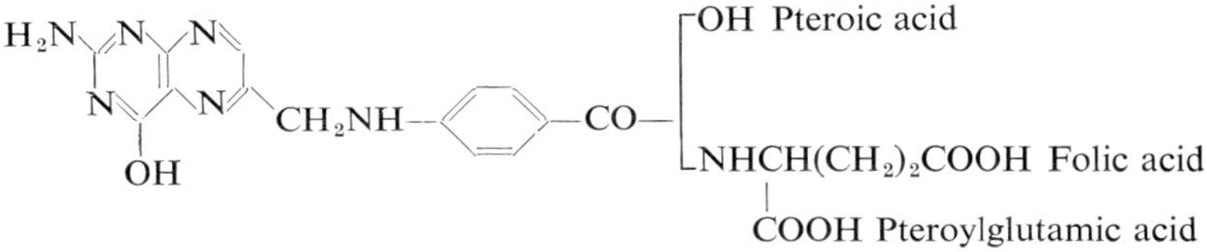

Examples of bacterial growth factors

cations in order to function properly. Examples of essential metals (anions) include Na^+, K^+, Cu^{++}, Mg^{++}, Ca^{++}, Zn^{++}, V^{++}, Mn^{++}, Fe^{++}, Co^{++} and essential cations include Cl^-, SO_4^{--}, again different species having different requirements for optimum growth.

In addition to the essential nutritional requirements listed above, there must be present at least one constituent which can be metabolised and which during metabolism provides energy for growth to occur. Examples of energy-yielding reactions will be considered in a later section.

It is possible to classify groups of bacteria on a basis which takes only the nature of their nutritional requirements and mode of obtaining energy into account. Two such schemes are given below.

NUTRITIONAL CLASSIFICATION OF MICRO-ORGANISMS

In the first place micro-organisms may be divided into two groups, autotrophs and heterotrophs, according to the nature of the carbon-containing derivatives they are able to utilise.

Autotrophic Bacteria

The autotrophs are distinguished by the fact that they can obtain their sole carbon requirements from an inorganic source such as a carbonate, bicarbonate or from carbon dioxide. They need only ammonia or nitrate as a source of nitrogen and may be further subdivided into two sub-groups based on the manner in which the individual members obtain energy for growth.

(*a*) *Chemosynthetic autotrophic bacteria.* This group can metabolise inorganic substances and thereby obtain energy for growth. Thus an organism occurring in the soil, *Nitrosomonas europaea*, can oxidise an ammonium salt to a nitrite and utilise oxygen from the atmosphere to effect this oxidation. *Thiobacillus thio-oxidans* can oxidise sulphur to sulphuric acid, producing energy for growth. This reaction is a potential nuisance in gas works, as by the action of this organism on the inert sulphur a corrosive product, sulphuric acid, is obtained.

(*b*) *Photosynthetic autotrophic bacteria.* Organisms in this group contain a pigment related to the chlorophylls of higher plants and oxidise inorganic substances, utilising the energy of sunlight.

Heterotrophic Bacteria

These organisms are unable to use bicarbonates or carbon dioxide as their sole carbon source but need a source such as glucose, lactic acid, glycerol, cellulose or starch. Energy for growth is obtained from the metabolism of these compounds. Small amounts of carbon dioxide are, however, required by the majority if not all heterotrophic organisms.

Within this group a further subdivision based on nutritional requirements may be made:

(*a*) Nitrogen-fixing organisms.
(*b*) Non-exacting organisms.
(*c*) Organisms requiring one or more pre-formed organic compounds before growth can occur.

(*a*) *Nitrogen-fixing organisms.* These organisms are of great economic importance in that they effect a natural fertilisation of the soil. The members of the genus *Azotobacter* are the most important nitrogen-fixing micro-organisms.

(*b*) *Non-exacting organisms.* These have no specific requirement for pre-formed organic molecules other than an organic compound which is capable of being metabolised and providing energy; a nitrogen source such as an ammonium salt or a nitrate, the ions and trace elements already listed and small amounts of carbon dioxide. Non-exacting organisms will grow on a simple synthetic medium of which the following is an example:

$(NH_4)_2HPO_4$	60 mg
KH_2PO_4	40 mg
NaCl	100 mg
$FeSO_47H_2O$	1 mg
$MgSO_4,7H_2O$	5 mg
Glucose	2 g
Water	to 100 ml

Good growth is obtained in this medium with, for example, *Escherichia coli*, *Serratia marcescens*, *Aerobacter aerogenes* and *Pseudemonas aeruginosa*.

These four organisms are thus examples of nutritionally non-exacting species and can synthesise proteins, lipids, nucleic acids and other cell constituents from the components of this medium.

It should not be overlooked, however, that if amino acids or growth factors or a natural source of growth factors, such as yeast extract, are added to a minimal medium, a greater rate of growth and a shorter generation time is usually obtained.

Another factor to be borne in mind is that even if chemicals of analytical reagent grade are used to prepare the medium, these chemicals will contain as impurity trace elements which may be in sufficient quantity to provide the necessary amount for growth, so that in this sense the medium is not completely defined.

(*c*) *Exacting organisms*. In this group can be found organisms which need a single growth factor or amino acid and others which need many. Thus *Salmonella typhosa* requires the amino acid tryptophan, while *Proteus vulgaris* requires the growth factor nicotinic acid. *Lactobacillus arabinosus* requires at least 24 preformed amino acids and growth factors. Further examples and details are given in Table 3.

In some texts a different terminology is used because the above classification is defective in its treatment of the nutrition of the photosynthetic bacteria some of which use organic compounds as a growth source.

In this terminology, four groups are recognised, photolithotrophs, photoorganotrophs, chemolithotrophs and chemoorganotrophs. The prefixes photo- and chemo- imply that the organism uses light or a chemical substance as an energy source; the terms -litho- and -organo- imply an inorganic or organic food source is used.

Organisms considered in detail in Chapter 4 belong either to the heterotrophic groups b and c of the first scheme or are, in the terminology of the second scheme, chemoorganotrophs.

The absolute requirement of micro-organisms for certain substances has been exploited for the analytical determination of these substances, many of them the vitamins implicated in human nutrition, by the process known as microbiological assay. In brief, it can be said that within certain limits the extent of bacterial growth is proportional to the amount of growth factor or amino acid present. By comparing growth obtained with known amounts of vitamin with that containing unknown amounts, the amount of vitamin, amino acid or other factor may be computed.

THE OXYGEN REQUIREMENTS OF BACTERIA

The bacteria may be divided into four groups by a consideration of their oxygen requirements.

Strict aerobes. These organisms have an absolute requirement for oxygen if they are to reproduce.

Strict anaerobes. Organisms which can multiply only in the complete absence of oxygen, the determining factor for growth being that the oxidation-reduction potential of the medium reaches a low value, characteristic for each culture. It is doubtful, however, if this is the only reason for the inhibitory or toxic effect of oxygen, and recently the theory that NAD and NADP (chapter 3) necessary for the metabolism of the organism, are used up to the organisms detriment in the unrewarding task of reducing molecular oxygen.

But as yet, there seems to be no unifying theory to explain an accumulating data.

Facultative organisms. Members of this group can grow either with or without oxygen, that is aerobically or anaerobically. Usually, all things being equal, better growth is obtained aerobically than anaerobically. This phenomenon is called the Pasteur effect.

Micro-aerophilic organisms. These have a requirement for oxygen, but show optimum growth at partial pressures of oxygen lower than encountered in the atmosphere.

Anaerobic Growth and Techniques for the Cultivation of Anaerobes

The cultivation of bacteria under anaerobic conditions requires special techniques; these may be divided into two classes: those methods which consist of placing the cultures in an air-tight container and removing all oxygen (*a–c*), the use of a medium in which oxygen diffuses with difficulty (*d*), and those methods which consist of adding a non-toxic reducing agent to the culture medium (*e–g*).

(*a*) *Boiling.* The culture medium is dispensed in bottles fitted with a screw cap with central aperture and a rubber washer; the cap is loosened, the bottle placed in a water-bath and brought to boiling and held for 10 minutes. The cap is tightened and the bottle allowed to cool. The organism is inoculated by means of a hypodermic syringe, the needle being plunged through the rubber washer.

(*b*) *Use of alkaline pyrogallol.* The cultures are placed in a desiccator containing alkaline pyrogallol to remove oxygen. A modification of this method consists of a petri dish with a double compartmented lid, one to contain sodium hydroxide solution, the other to contain pyrogallol solution. The organism to be grown is streaked on to a suitable agar, placed agar downwards over the lid and the two solutions mixed by gentle agitation. The assembly is then incubated.

(*c*) *The McIntosh and Fildes jar*. This consists of a jar constructed of strong glass, fitted with a gas-tight metal lid which can be securely clamped in place. This lid carries two terminals which convey an electric current to a coil within the jar which will heat a piece of spongy palladium deposited on asbestos, and surrounded with wire gauze. The lid is also fitted with inlet and outlet taps. To use the apparatus the cultures are placed in the jar and hydrogen passed through, via the taps; the current is switched on which heats the spongy palladium; oxygen present combines with hydrogen in the presence of the heated catalyst; the reaction should be prevented from reaching explosive violence by the external gauze, but it is normal to place the jar in a protective box when carrying out this part of the operation. Hydrogen is passed for a further 30 minutes and the taps are then shut and the jar incubated at a suitable temperature. An indicator of anaerobiosis may be included in the jar and consists of a mixture of methylene blue and glucose made alkaline with sodium hydroxide. This is heated, when the alkaline glucose solution decolorises the methylene blue. The decolorised solution is placed in the jar and should remain colourless during incubation period—any reappearance of the blue colour is indicative of leakage and is caused by the oxygen in the atmosphere. Catalysts which do not require heat are now widely used.

(*d*) *Growth in solid and semi-solid media*. If solid medium is used anaerobic conditions may obtain if this medium is poured into a test tube and the organisms inoculated near to the bottom with a long wire. This is known as a stab culture. Oxygen is eliminated in the sense that it is boiled off when the medium is sterilised in the tube and is prevented from reaching the bottom of the tube by diffusion, owing to the height and nature of the layer of medium above.

A further modification of this type of culture is to use a nutrient broth containing just enough agar to render it semi-solid—the so-called sloppy agar. This medium is heated and cooled just before inoculation. Anaerobic conditions should be maintained at the bottom of the container, the viscosity of the medium preventing oxygen access by diffusion.

(*e*) *Metallic iron*, in the form of wire or small iron nails, is effective in producing anaerobiosis. The medium is heated in a boiling water-bath for 10 minutes, the iron sterilised by heating in a bunsen flame for 45 seconds, then added to the medium. The normal aerobic medium used to study decomposition of sugars by aerobes and facultative organisms can be thus adapted to studying their decomposition by anaerobic organisms.

(*f*) *Cooked meat.* Sufficient minced, cooked, ox-heart muscle is added to a conventional nutrient broth dispensed in a screw-capped container of 30 ml capacity so as to form a 1·5 cm deep layer at the bottom of the liquid. Muscle tissue contains reducing systems. One system consists of the reduction brought about by the irreversible oxidation of lipids, the other the reversible para-haematin-haemochromogen system. This medium was originated by Robertson and is often known as Robertson's cooked meat medium.

(*g*) *Thioglycollic acid.* This substance, $CH_2SHCOOH$, has been used successfully as an additive to nutrient broths for the cultivation and detection of anaerobes since 1926. Sodium sulphide had been used for the same purpose eight years earlier. A medium containing thioglycollic acid and other ingredients was introduced by Brewer in 1940 for the purpose of detecting both aerobes and anaerobes primarily for use in testing certain pharmaceutical products for absence of bacterial contamination (sterility tests). Since the original recipe by Brewer, other formulae have been tried; a typical example is given below:

Meat extract	1·0 g
Yeast extract	2·0 g
Peptone	5·0 g
Glucose	5·0 g
Sodium chloride	5·0 g
Sodium thioglycollate	1·0 g
Methylene blue	0·002 g
Agar	0·5 g
Water	to 1,000 ml

This consists of a nutrient broth (first five ingredients), the reducing agent (thioglycollic acid), an indicator of the extent of reduction (methylene blue) and sufficient agar, 0·05 per cent, to increase the viscosity of the medium and reduce the tendency for air to diffuse into the medium. The medium is heated before use; during this time the methylene blue is reduced to leuco-methylene blue and the medium appears its natural yellow colour; on oxidation the leuco-methylene blue becomes blue and the overall appearance of the medium is green. There is thus at any time a visible indication of the extent of reductive conditions in the tube or container of medium.

Oxidation-Reduction Potentials of Bacterial Cultures and Anaerobic Growth

It is apparent from the information given above that there is a relationship between anaerobic growth and the level of oxidation or reduction in the environment. The quantitative determination of this level is made by measuring a factor known as the oxidation-reduction or redox potential. This may be likened to the pH value which is a measure of the degree of acidity or of alkalinity.

A full theoretical treatment of oxidation-reduction potentials will be found in textbooks of physical chemistry. Essentially, however, the process of oxido-reduction may be represented thus:

$$\text{Reduced form} \underset{\text{reduction}}{\overset{\text{oxidation}}{\rightleftharpoons}} \text{oxidised form} + n \text{ electrons}$$

This process should be compared with the ionisation of hydrogen

$$\text{H} \rightleftharpoons \text{H}^{+} + 1 \text{ electron}$$

Both these reversible reactions result in the appearance or disappearance of electrons. If an electrode consisting, for example, of a 5 mm square piece of platinum is placed in a solution containing a redox system or systems, a potential is established at that electrode. By connecting this electrode via a very sensitive voltmeter (usually a potentiometer containing an amplifying circuit) to a second electrode of known potential, the potential of the inert electrode may be evaluated.

In practice the measuring instrument is calibrated in terms of volts or millivolts to enable direct readings to be made.

It can be shown that, at constant pH, the following relationship holds:

$$E_h = E_0 + \frac{RT}{nF} \log_e \frac{[\text{Ox}]}{[\text{Red}]}$$

where E_h = the measured electrode potential in volts with respect to the hydrogen scale

E_0 = the value of the electrode potential in volts at 50 per cent oxidation

R = gas constant (8·3 absolute joules/degree/mole)

T = temperature of the system expressed in degrees absolute
F = the faraday, 96,500 coulombs
n = the number of electrons involved in the reaction
[Ox] = the concentration of the oxidised form in gm-ions/litre
[Red] = the concentration of the reduced form in gm-ions/litre

If [Ox] = [Red], that is a condition in which the system has undergone 50 per cent of the theoretical total oxidation, [Ox]/[Red] = 1, therefore $\log_e$ [Ox]/[Red] = 0, thus $E_h = E_0$. E_0, then, is the value of the electrode potential at 50 per cent oxidation, or 50 per cent reduction, and is called the standard electrode potential for the system; it is a measure of the proneness of a system to yield electrons, that is to undergo oxidation just as pH is a measure of the proneness of an acid to yield hydrogen ions. The practical meaning of a scale of E_0 values is that any redox system will oxidise a second redox system of lower E_0 value, or reduce a system of higher E_0 value. As might be expected, E_0 is dependent on the pH of the surrounding solution; the equation is written for a pH value of 0. It is customary to give values of E at a constant pH, usually 7, when the symbol E' is used, the pH being stated.

Values for E_h may also be determined with appropriate dyestuffs. Thus, at pH 7·0, methylene blue (used in the anaerobic medium mentioned above) is fully oxidised at E_h' of +0·5 volts, fully reduced at an E_h' of −0·5 volts, that is about 50 per cent reduced at E_h' of 0·0 volts. There are available a whole range of dyes with different E_0' values for colorimetric determination of redox potentials (Table 4).

Interesting results are obtained if the redox potentials of growing cultures of bacteria are measured. The E_h value of nutrient broth is of the order of +0·03 volts. If this broth is inoculated with an obligate aerobe the E_h value falls to between −0·15 and −0·20 volts.

The E_h value of nutrient broth containing cooked heart muscle, or of thioglycollate broth, is about −0·20 volts, due to the presence of the reducing agents. Anaerobic organisms will grow in this medium and reduce the potential still further, to about −0·4 volts.

Two facts emerge from these data; firstly that bacteria tend to reduce the redox value of media in which they are growing, and

TABLE 3

Examples of redox systems	E_0' pH 7 and 30° (volts)
Natural systems	
Cytochrome *a* ⇌ reduced cytochrome *a*	+0·29
Cytochrome *c* ⇌ reduced cytochrome *c*	+0·25
Cytochrome *b* ⇌ reduced cytochrome *b*	+0·002
ubiquinone⇌reduced ubiquinone	−0·10
Flavin-adenine dinucleotide ⇌ reduced flavin-adenine dinucleotide	−0·06
Nicotinamide-adenine dinucleotide ⇌ reduced nicotinamide-adenine dinucleotide	−0·32
Nicotinamide-adenine dinucleotide phosphate ⇌ reduced nicotinamide-adenine dinucleotide phosphate	−0·32
Ferredoxin ⇌ reduced ferredoxin	−0·42
Redox dyes	
Benzylviologen	−0·36
2,6-Dichlorophenol indophenol	+0·217
Methylene blue	+0·011

secondly there is a differing scale of values over which aerobes and anaerobes will grow. In fact, studies with many anaerobes have shown that the fundamental requirement for their growth is that the E_h of the medium must be at or near −0·2 volts. Exclusion of oxygen or the presence of reducing agents promotes this condition.

The first observation may be explained by the fact that during the intense metabolic activity attendant upon growth (Chapter 3), many substances are oxidised to provide energy for growth; at the same time other substances will be reduced. These substances may range from dissolved oxygen to less obvious reducible substances, present in the medium. The production of these reduced substances accounts for the fall in potential observed during metabolism.

A partial explanation of the second observation may lie in the E_0' value of the oxidation-reduction enzymes taking part in the metabolic processes of these two groups of micro-organisms (Table 4). Anaerobes contain the redox system ferredoxin ⇆ reduced ferredoxin ($E_0' = 0{\cdot}42$ volts) discovered in 1962 and this low potential redox system clearly operates in the E_0' region in which anaerobes are known to flourish and has been shown to be part of the respiration mechanism of these organisms. In contrast nearly all aerobes utilise in their terminal metabolism a group of pigments called the cytochromes (page 84), which are characteristically absent from the anaerobes. In this family of respiratory pigments cytochrome *a*, which reacts with and reduces oxygen, has an E_0' value of +0·29 volts, which is of the same order as that of normal nutrient broth, +0·30 volts. The aerobic metabolism of the streptococci (page 132), devoid of cytochromes may be explained by the fact that the H_2O_2 produced is decomposed by peroxidases.

It is sometimes assumed that when redox potentials of cultures are being determined the redox potential of the bacterial cell is being measured. This assumption is not warranted. The observed changes must measure the overall changes in the bacterial culture brought about by the metabolic activity of the growing cells.

THE EFFECT OF pH ON BACTERIAL GROWTH AND VIABILITY

The pH of the immediate environment has a significant effect on bacterial growth. There is for a given set of conditions an optimum pH and at the extremes of the pH scale growth may be prevented altogether and finally the inoculum may be destroyed.

It is important to bear in mind that pH differences can affect both the nature of the medium as well as having a direct effect on the micro-organism. If the medium contains ionisable constituents, such as amino acids or salts of weak acids or bases, the utilisability of these compounds by the cell may be affected and this may be a contributory cause of growth inhibition.

The effect of hydrogen ion concentration upon the bacterial cell will first manifest itself on the bacterial surface; this structure contains ionogenic groups and most bacteria bear a net negative charge at pH 7·2–7·6. Changes in the pH of the environment change the degree of ionisation of cell surface components and this factor must contribute to the inhibitory effect of extremes of pH. Eventually the cytoplasm with its complement of enzymes and nuclear material will be affected.

At pH values about 1·5–2 units nearer neutrality than the lethal pH, a state of bacteriostasis may occur. Thus it has been shown that 36 hours at pH 2·5 failed to sterilise an inoculum of 10^6 cells/ml of *Esch. coli* at 20°, whereas at pH 1·5 the inoculum was sterilised in 15 minutes. At pH 11 the same inoculum was sterilised in 10 minutes.

If a plot is made of the growth (if any) of a micro-organism against the pH of the medium (sometimes referred to as the bulk pH), a curve is usually obtained of the type as shown in Fig. 14. This exhibits for the particular experimental conditions a maximum, termed the pH optimum, and regions of no growth where the inhibitory or lethal process of extremes of pH have occurred—

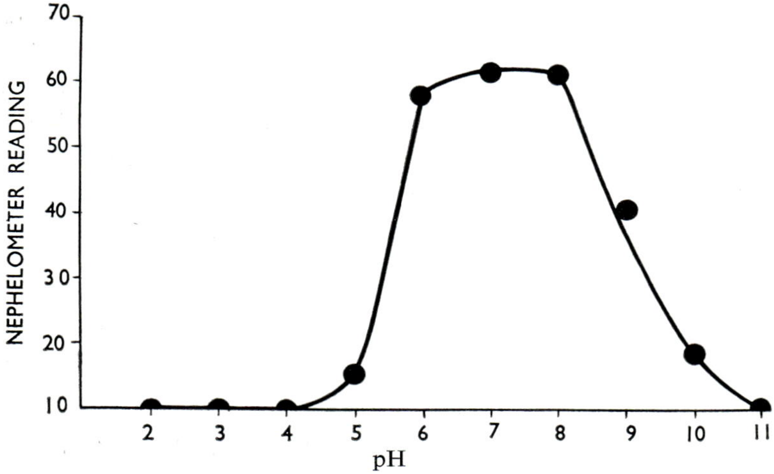

FIG. 14. Effect of pH on bacterial growth. *Escherichia coli* in nutrient broth. (W. B. Hugo)

and represents the integration of all the various effects of pH on cell growth.

It is important to make sure during an experiment of this kind, usually carried out in a buffered nutrient medium, that significant pH changes are not caused by the metabolic activity of the organism, thereby vitiating the results.

Use has been made of the bacteriostatic and bactericidal effect of extremes of pH in sterilisation and food preservation. Vinegar is used to preserve vegetables, while quick-lime has been used to sterilise infected animal carcases.

Moulds are in general more tolerant of acid conditions (pH 5–6) than are bacteria. By adjusting media for the growth of moulds to the acid side of neutrality, the likelihood of growth of bacteria is reduced.

In nutrient broth the optimum pH for most human pathogenic bacteria has been found to be slightly on the alkaline side of neutrality. The medium used for official tests for sterility is adjusted to pH 7·4–7·8.

A mechanism which may operate as an *in vivo* pH controlling device is described on page 108.

THE EFFECT OF TEMPERATURE ON BACTERIAL GROWTH AND VIABILITY

It is well known that increasing the temperature at which a chemical reaction is performed will increase its rate. A supply of energy in the form of heat to the reactants may be necessary to start a chemical reaction, and thereafter maintain it.

All activities of bacteria are essentially chemical reactions, involving the decomposition (catabolism) or synthesis (anabolism) of new compounds. These reactions are diverse and include the synthesis of new cellular material, of complex pigments or of antibiotics to quote only a few examples. One cardinal feature stands out concerning chemical reactions carried out by bacteria, they are catalysed by enzymes which are protein in nature or contain a pro-

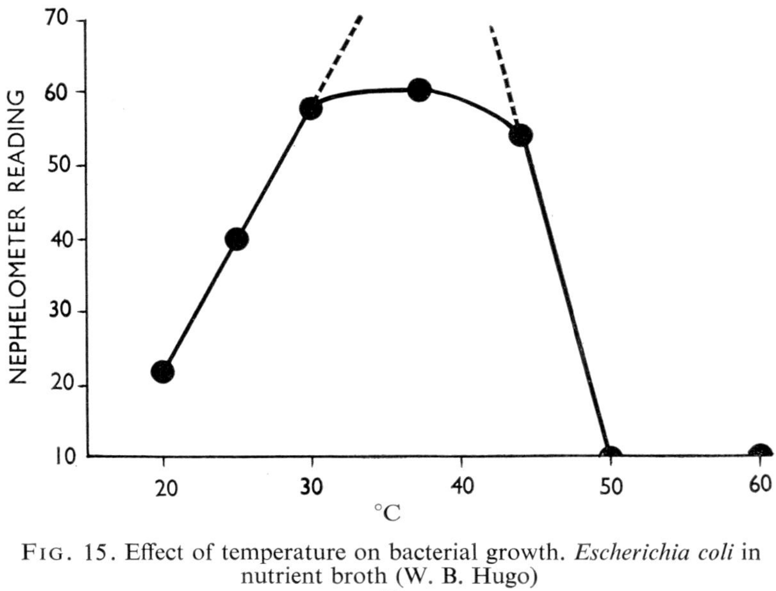

FIG. 15. Effect of temperature on bacterial growth. *Escherichia coli* in nutrient broth (W. B. Hugo)

tein moiety. Proteins are destroyed by heat at relatively low temperatures. Thus there is competition between the two factors, the normal effect of increased temperature on reaction rates and the destructive effect of heat on proteins.

A study of the effect of varying temperatures on the growth of bacteria (a process which represents the integrated results of very many chemical reactions) produces the typical bell-shaped curve already encountered with the pH growth relationship. This curve consists of two components, A and B (Fig. 15). Component A represents the increase in rate of chemical reaction (as represented by growth) with increasing temperature. Component B represents the slowing of the overall processes contributing to growth. At first the slowing may only be due to the breaking of the co-ordinated series of reactions necessary for optimum growth because of the inability of one enzyme in a key reaction to function smoothly in co-ordination with others at slightly above the optimum temperature. With increase in temperature, however, the destruction of enzyme protein and other cell constituents begins and finally the cells are killed.

A survey of the temperatures at which bacteria can grow and their temperature optima has revealed an interesting pattern that has enabled bacteria to be divided into three classes, called psychrophiles or psychrophilic bacteria, mesophiles or mesophilic bacteria and thermophiles or thermophilic bacteria.

Psychrophilic bacteria. These organisms grow best at 15° to 20° C, can grow at 0° to 5° C and do not grow above 30° C. It can be seen that they will proliferate at refrigerator and cold storage room temperature, and bacteria in this group have been responsible for spoilage in such commodities as refrigerated poultry carcases, their metabolic products causing taint and discoloration. Under deep-freeze conditions, −20° to −40° C, the growth of psychrophiles is suppressed.

Mesophilic bacteria. These organisms grow best between 25° and 40° C. They do not grow above 45° or below 5° C. The organisms pathogenic to man and animals and many non-pathogenic species found in the soil and water fall into this group.

Thermophilic bacteria. The optimum temperature of this group is of the order of 55° C; some will grow at 80°, while little or no growth occurs below 25° C. Organisms in this group are responsible for one type of spoilage in heat-preserved canned foodstuffs.

The Use of Heat as a Sterilising Agent

Heat sterilisation is used extensively in the food and pharmaceutical industry and in medical and surgical practice. Two methods of utilising heat as a sterilising agent may be employed—moist heat and dry heat. The former consists of applying heat in the presence of water either as a liquid or vapour. Dry heat refers to heating in a non-aqueous environment such as a hot-air oven.

Flaming is a process of dry heat in which the object to be sterilised is heated directly in a flame. It is of limited application but is used routinely to sterilise inoculating loops and wires. Contaminants are charred and oxidised probably to carbon dioxide and water.

Moist heat is much more effective than dry heat for the destruction of both vegetative bacteria and spores. Thus moist heat at 100° C for 5 minutes will kill all vegetative bacteria, but 90 minutes may be necessary at the same temperature in the absence of extraneous moisture. For spores, in general, 115° C for 30 minutes is lethal in the presence of moisture; in the absence of extraneous moisture a temperature of 150° to 160° C maintained for 60 minutes may be necessary. Thus it is the ability to destroy bacterial spores which is the significant criterion of the efficacy of a sterilisation process.

Reports appear in the literature of bacterial spores with unusual heat resistance. Thus the spores of some strains of *B. anthracis* may require 120 minutes at 160° to render them non-viable. Outstanding in their resistance to heat are the spores of thermophilic organisms. These organisms do not include pathogens but are a cause of spoilage in food products and if detected in a sterility test could cause a batch of pharmaceutical products to be rejected for use.

The technology of sterilisation is described in a separate monograph in this series (Vol. 3).

Thermal Death Rate and Thermal Death Time

These two values are of practical use in formulating sterilising procedures both for the food and pharmaceutical industries and in evolving the official methods of sterilisation involving the use of heat. To determine the thermal death rate of an organism, it is exposed for a constant time at varying temperatures and its viability determined by a suitable method.

The thermal death time is found by determining the time to kill all the organisms in an inoculum at a constant temperature. If in the preparation of a heat-killed bacterial vaccine it is known that to ensure an active preparation the organism must not be heated above 65° C, then the time to achieve complete kill of all organisms at this temperature must be determined. Similar problems arise in the food industry; it may be impossible to heat certain foods above a certain temperature without causing adverse changes. The length of time of heating may be the important factor in determining a heat-processing regime.

The Decimal Reduction Time

In many texts, especially those dealing with food preservation and the destruction of bacterial spores by radiant energy (page 65) a factor called the D_{10} or decimal reduction time is used to quantify the lethal process. It is defined as the time in minutes to produce a 90 per cent kill.

Although widely used, it is really only of help in assessing the general reduction of viability and is in no way a measure of sterilisation. Indeed it might well be from our knowledge of the kinetics of bacterial death (Vol. 3, p. 222–224) that the residual 10 per cent of organisms not included in the D_{10} assessment take a proportionately much longer time to die representing perhaps the residual resistant members of the population.

The Mechanism of the Lethal Effect of Heat

The marked difference in the temperature and time required to effect sterilisation by moist and dry heat suggests that different mechanisms of kill may be involved.

The mechanism of thermal destruction by moist heat probably involves extensive hydrolysis of cell components; for example, forces responsible for holding protein molecules in their characteristic folded or coiled shape may be broken so that unfolding of the protein chain occurs (protein denaturation). Other macromolecular components may also be affected. The lethal effect of dry heat is thought to be largely due to the oxidation of cellular components. Lipids and the thiol (—SH) groups which occur in many proteins and in coenzyme A are particularly prone to oxidation.

The Effect of Cold on Bacteria

Extremes of cold are usually not lethal and vegetative bacteria have been known to survive 20 hours' exposure to liquid air

($-185°$ C). In one experiment the temperature of liquid hydrogen ($-250°$ C) was found not to be lethal after 10 hours for several vegetative organisms and *B. anthracis.*

Alternative freezing and thawing, however, is a lethal process and was used as a method of extracting intracellular material from bacteria at the beginning of this century. The lethal effect of this process is probably due to the mechanical rupture of cells by local freezing of parts of the cytoplasm and local changes in internal pressures.

Because of the general slowing down of bacterial growth and metabolism, refrigeration and deep freezing form a useful method of food preservation.

THE EFFECT OF DESICCATION

Many bacteria survive drying, although the ability to do so differs widely from species to species, and is dependent on factors such as the rate of drying and the nature of the environment in which the drying took place. For example the presence of colloidal matter such as serum during the drying process renders it less lethal. Bacterial spores are even more resistant to drying than vegetative forms and survival periods of the order of a quarter of a century have been recorded for the spores of some species.

Slow drying of bacteria under high vacuum, frozen with a colloidal substance, a process known as freeze-drying or lyophilisation, is now an accepted and widely used method of preserving bacteria for stock cultures in laboratories and culture collections.

THE EFFECT OF OSMOTIC PRESSURE ON BACTERIAL GROWTH

It will be remembered that the bacterial cell consists of a protoplast with an internal osmotic pressure of 6–25 atmospheres surrounded by a rigid cell wall. The protoplast membrane exerts a controlling influence over the outward diffusion of solutes and the inward diffusion of water. Bacteria, in general, are not affected by being placed in distilled water or in dilute solutions of solutes, although a sudden transfer from a medium of near the maximum tolerable osmotic pressure to distilled water may cause the cell to burst (plasmoptysis). This is due to a rapid uptake of solvent into the cell and a sudden change from distilled water to a solution containing 2–5 per cent of solute such as sodium chloride may result in a shrinkage of the protoplast away from the rigid cell wall (plasmolysis). At higher salt concentrations, growth is inhibited. Typical figures for the growth inhibitory activity of sodium chloride are 7·5–20 per cent, dependent on the bacterial species.

The use of brine for preserving meat depends on the inhibitory effect of near-saturated aqueous sodium chloride. It is salutary to note, however, that bacteria capable of causing spoilage in meats have been isolated from a liquid environment saturated with sodium chloride. Even concentrated solutions such as pharmaceutical syrups, concentrated fruit juices and honey are liable to decomposition by micro-organisms, especially yeasts, while the sea (3·5 per cent total salinity) contains its own microbial population and is not lethal to some human intestinal bacteria that reach it via sewage systems.

An interesting use is made of the differing susceptibilities of food-poisoning organisms to sodium chloride. It was found that pathogenic staphylococci could grow in the presence of 7·5 per cent sodium chloride in a solid medium, whereas non-pathogenic staphylococci grew less well and most other bacteria would not grow on such a medium. A nutrient agar containing 7·5 per cent sodium chloride is used as a selective medium in the isolation of staphylococci.

WATER AND BACTERIAL GROWTH

On page 38 water was indicated, 6, as an essential ingredient for microbial growth and immediately above, osmotic pressure was shown also to have an effect on growth.

It is possible to quantify water requirement and osmotic pressure by the use of a factor called available water or water activity, A_{ω}.

Water exerts a finite vapour pressure, P_0, at constant temperature and pressure; if substances, whether as pure chemical entities such as sodium chloride or complex substances such as peptones are dissolved in Water, the vapour pressure, P, exerted is lowered.

The ratio P/P_0 is equal to a_{ω}, the water activity, and this dimensionless factor, always a fraction of unity, when multiplied by 100 gives the relative humidity of the solution, i.e. relative to the 'humidity' of pure water again at the same temperature and pressure.

It will further be recalled that the ratio P/P_0 gives, when multiplied by suitable constants, the osmotic pressure of a solution, this value being a dimensioned quantity. In summary therefore we have

$$\frac{P}{P_0} = a_\omega = \frac{\text{relative humidity}}{100} = \text{function of the osmotic pressure.}$$

When actual figures are examined it is found that most bacteria grow between a_ω values of 0·995 to 0·990. However, *Staph. aureus* will grow at a_ω values 0·86. This illustrates a quantification of the statement made in the section above concerning the uses of a media containing 7·5 per cent sodium chloride to select staphylococci. Some of the salt tolerant, halophilic, organisms isolated from brines can grow at a_ω 0·75. Moulds and yeasts often show growth at even lower values.

THE EFFECT OF IRRADIATION ON BACTERIAL GROWTH

It has long been known that exposure of bacterial cultures to direct sunlight has an inhibitory effect on growth, and bacterial cultures are usually incubated in the dark; experiments have shown that light in the ultraviolet region, of wave-lengths from 200 to 280 nm, are bactericidal. Gamma radiation (short X-rays) of wave-length 0·005–1·0 nm, particulate radiation, such as a high-speed electron beam (β-rays), helium nuclei (α-rays) or neutron beams are also lethal. Irradiation techniques are used in sterilisation procedures.

The Mechanism of the Lethal Effect of Radiations

It is convenient to distinguish two types of effect which radiations can exert upon the cell. The first is a direct effect on cell constituents, the second is an effect due to the action of toxic radicals which are produced usually from water in the environment or within the cell.

The effect on cell constituents. It is well known that radiations of wave-lengths varying from the ultraviolet to the microwave range are absorbed by molecules and atoms and upon this fact depends the science of spectroscopy. In certain instances the absorbence of radiant energy results in a permanent or irreversible change in a molecule. Particulate radiations may also permanently damage cellular constituents.

All molecules in the cell may be affected but damage to one such molecule, deoxyribonucleic acid (DNA) (page 110), is of particular significance.

DNA molecules in the nuclear apparatus of micro-organisms are responsible for the inheritable characters of the organism (page 15). A reversible chemical change in a gene may result in the loss of a single characteristic, for example the ability to synthesise an essential amino acid. It is clear that if that amino acid is not available in the environment, the affected cell will die. It is not difficult to envisage a number of these events or mutations occurring if a cell population is irradiated, thus resulting in the death of the population; such changes are called lethal mutations. Not all mutations are lethal, however, and irradiation is used in the selected breeding of strains of micro-organisms. Thus by the treatment of the original (wild) *Penicillium chrysogenum* with X-rays, a strain which did not produce the characteristic yellow pigment was produced, resulting in the production of penicillin uncontaminated with the yellow pigment.

Recently a detailed mechanism for DNA damage by ultraviolet light has been discovered. Thymine, one of the bases in DNA (page 15), will dimerise with another adjacent thymine molecule in the DNA chain when irradiated with ultraviolet light of wave-length 200 nm. This event, it is thought, is sufficient to impair or prevent the biological function of a DNA molecule.

Photoreactivation. It has been known for some time that irradiation of ultraviolet-damaged bacterial cells with visible light, especially blue light, will restore much of the damage, even apparently reviving cells which otherwise might have been expected to die. This is known as photoreactivation. An explanation of this phenomenon has recently been advanced. An enzyme has been shown to exist in bacteria, activated by light, especially blue light, which can reverse the formation of dimerised thymine molecules. Some bacteria contain an enzyme similar in its action on breaking thymine dimers but which does not require light for activation. Such bacteria are, not unexpectedly, resistant to ultraviolet light.

This theory of the formation and destruction of thymine dimers with its attendant deactivation and reactivation of DNA explains the target theory of ultraviolet radiation. This held that there is a vital target in the cell which if hit by the rays concerned is destroyed, consequent upon which the cell dies. The target could be, in many cases, the thymine molecule in DNA. The existence of a dimer-reversing enzyme also explains the phenomenon of photo-reactivation and the emergence of ultraviolet resistant strains of micro-organisms.

The DNA molecule, however, is not the only molecule to be affected and it is possible and indeed probable that some cells may be killed by the destruction of an essential enzyme, or other essential molecules.

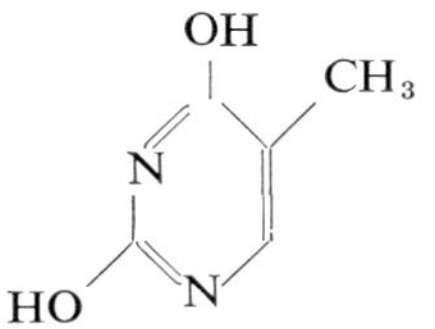

Thymine
2,4-Dihydroxy-5-methylpyrimidine

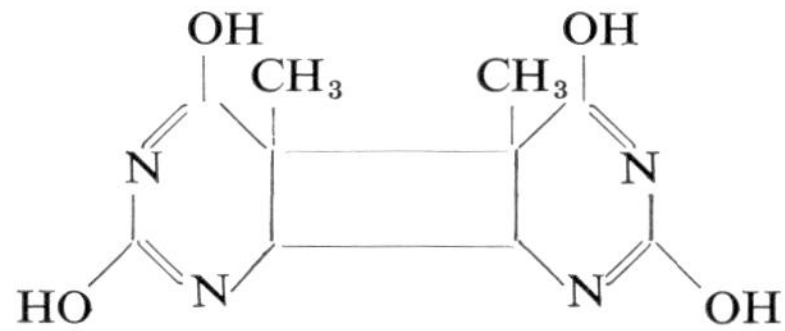

Thymine dimer

Secondary effects of radiation. Toxic-free radicals, as for example $HO_2\cdot$ and $HO\cdot$, are produced from water, either intracellular water or water in the environment following irradiation. Ionisation is a particular feature of X- and γ-ray and of neutron beam irradiation. These radicals are potent oxidising agents, and could irreversibly damage enzyme proteins containing thiol (—SH) groups which are particularly prone to oxidation. Evidence that oxidising agents play some part in the lethal effect is afforded by the experimental observation that the addition of readily oxidisable substances such as cysteine or sodium dithionite to liquid cultures reduces the lethal effect by reacting with the toxic oxidising radical. In many circumstances both mechanisms must play their part in the overall lethal effect of radiations and the logarithmic death curve is a fortuitous experimental integration of the several lethal events imposed upon the cell.

INHIBITION OF BACTERIAL GROWTH BY CHEMICALS

Most chemical substances other than those that are utilisable as food inhibit to a greater or lesser extent the growth of micro-organisms, and may in fact be lethal. Use has been made of this fact in evolving drugs for the treatment of infections due to all types of micro-organisms in man and animals and to rendering equipment free from micro-organisms, a very necessary procedure in the food industry and in the practice of medicine.

So important is this group of chemicals that very much thought and research has been devoted to methods of evaluating them and this aspect of the subject is dealt with in Volume 3.

Antimicrobial chemicals may in the first place be grouped into four classes. These classes are to a certain extent artificial and depend more on the general use to which the members are put than to a chemical delineation.

1. Chemicals used in fundamental metabolic studies.
2. Metabolite analogues.
3. Antiseptics or disinfectants.
4. Antibiotics.

Chemicals Used as Inhibitors of Specific Metabolic Stages

A small group of chemicals are used in studies involving metabolic pathways because of specific poisoning effects which they exhibit, and are dealt with in Chapter 3 (page 115).

Chemicals as Metabolite Analogues

Some chemicals bear such a close relationship to an essential metabolite that they become incorporated into the reaction sequence of the cell, but thereafter block further biosynthetic processes. If the kinetics of the enzymic processes involved are investigated, the metabolite antagonist is found to inhibit the process competitively (page 79, Fig. 18). Malonate has been known to be a competitive inhibitor for succinate since 1928. The antibacterial activity of the sulphonamides is due to the fact that they compete with *p*-aminobenzoic acid, a constituent of the folic acids, themselves essential for the growth of certain micro-organisms.

In addition to malonate and *p*-aminobenzoate many other antimetabolites or metabolic antagonists have been described. Examples are given in Table 4. Not all antimetabolites are of use in medicine for treating bacterial infection; they may prove to be as toxic to man or animals as to the invading bacteria. The sulphonamides owe their selective toxicity to the fact that man has an absolute requirement for the folic acids, that is he depends entirely on an external source for them and has no internal synthetic route to be damaged by the drug.

As with all classification schemes it is not possible to draw a clear line of demarcation between groups and it should be borne in mind that the use and study of antimetabolites has proved a useful aid in fundamental studies.

TABLE 4

Metabolite	Antimetabolite
Succinic acid CH_2COOH–CH_2COOH	Malonic acid $CH_2(COOH)_2$
L-alanine	D-alanine
p-Aminobenzoic acid 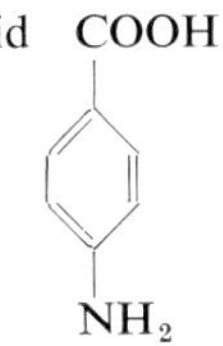	*p*-Aminobenzenesulphonamide (sulphanilamide) 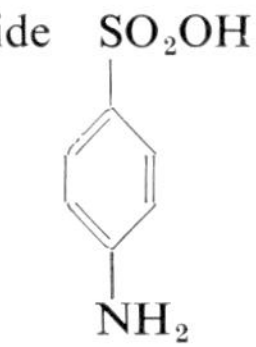
Nicotinic acid 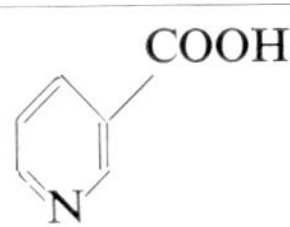 Nicotinamide 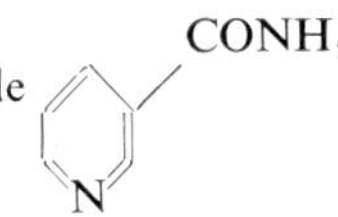	Pyridine-3-sulphonic acid 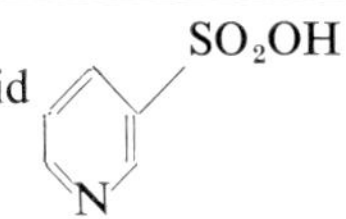
Thiamine 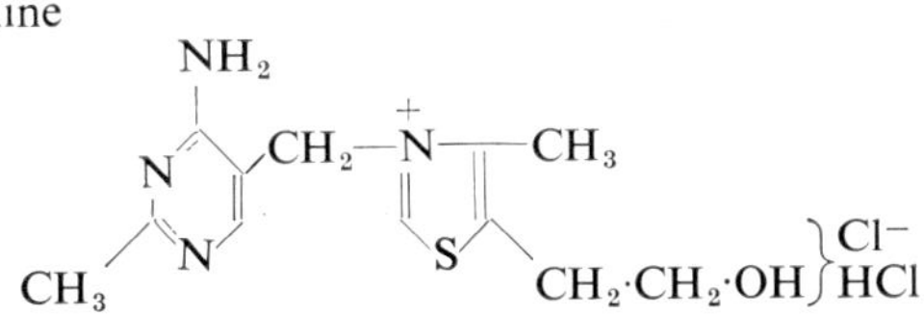	Pyrithiamine 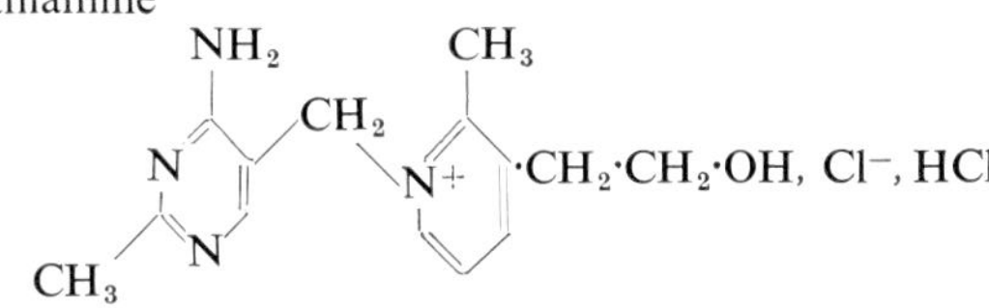

Antiseptics and Disinfectants

This group includes substances that may be used for the sterilisation of utensils or for application to wounds. Typical examples are phenols, and halogenated phenols, dyes, surface active compounds, halogens especially derivatives of iodine and chlorine, amidines, ethyl alcohol, formaldehyde solution, metal salts and derivatives of nitrofuran (Volume 3). In addition gaseous substances such as ethylene oxide and formaldehyde are used in sterilisation procedures.

Antibiotics

There is no special reason why antibiotics should be in a separate group except to say that they are antibacterial agents which are derived themselves from micro-organisms. Here again with the production entirely by chemical methods of the antibiotic chloramphenicol and the production of semisynthetic penicillins this definition is no longer rigid.

References

The Bacteria. Vol. 4: *The Physiology of Growth.* Ed. Gunsalus, I. C., and Stanier, R. V. Academic Press, 1962

The growth of bacterial cultures. Monod, J. Ann. Rev. Microbiol., 1949, **3**, 371

The Inhibition and Destruction of the Microbial Cell. Ed. Hugo, W. B. Academic Press, 1971

The Mode of Action of Antiseptics. Hugo, W. B. *J. appl. Bact.* 1967, 30, 17

Biochemistry of Antimicrobial Action. Franklin, T. J. and Snow, G. A. Chapman and Hall, 1971

Growth, Function and Regulation in Bacterial Cells. Dean, A. R. C., and Hinshelwood, Sir Cyril. Oxford University Press, 1966

Oxygen and Clostridia. Morris, J. G. and O'Brien, R. W. in *Spore Research 1971.* Eds. Barker, A. N., Gould, G. W. and Wolf, J. Academic Press, 1971.

CHAPTER 3

BACTERIAL METABOLISM

EARLY STUDIES IN THE BREAKDOWN OF COMPLEX MOLECULES

The formation of alcoholic beverages from a variety of natural products, notably grape juice and starches, has long been known to groups of mankind scattered widely over the earth. The parallel phenomenon of putrefaction, the breakdown of proteins, was similarly well known and both consist of the breakdown of complex molecules to give much simpler and often volatile products.

The elucidation of the nature of these two phenomena gave rise to a classical controversy in the scientific thinking of the eighteenth and nineteenth centuries. This controversy centred around two theories: one held that the spontaneous generation of micro-organisms in fermenting and putrifying systems could occur; the other held that there existed in nature a form of life, of small size and not apparently obeying the same laws of reproduction as seen in the more familiar living organisms on the earth, and if this form of life fell fortuitously upon matter it gave rise to fermentation or putrefaction. The Dutch amateur lens maker Antonie van Leeuwenhoek had in the 1680's seen and described such minute bodies in fermenting systems and other sources and was the first person believed to have seen bacteria.

During the middle of the eighteenth century a long dispute, backed by a variety of experiments, developed between Needham, a supporter of the theory of spontaneous generation, and Spallanzani, who believed in the microbial theory of decomposition of natural products. Many of the results of the experiments and counter-experiments performed by these two early scientific workers and others would have been explicable if the existence and properties of the bacterial spore had been known and understood.

A century later, both theories still had their protagonists with the weight of belief in favour of the theory of spontaneous generation, despite the fact that Cagniard-Latour had in 1838 stated his conviction that fermentation was due to the growth of yeast in sugar solutions, and in the same year Schwann had independently stated the same view. The last contestants to enter the field were Berzelius and Liebig, who held an essentially chemical view of the fermentation and putrefaction process, and Pasteur, who intuitively believed the theories of Cagniard-Latour and Schwann, that is, that a living organism was implicated.

Berzelius contended that fermentation was due to the presence of a non-living catalyst or ferment, which was ubiquitous, while Liebig believed that the decomposition of carbohydrates was dependent on the spontaneous co-decomposition of dead animal or vegetable matter, the latter producing ferments or agents of decomposition. In the last analysis, Liebig's theory was found to be untenable, but although Pasteur believed that fermentation depended on the presence of living cells, thus apparently contradicting Berzelius, a year after Pasteur's death, which occurred in 1895, Buchner made a cell-free juice from yeast which was capable of initiating fermentation. Thus, at last, the apparently opposing ideas of Berzelius and Pasteur were reconciled because in fact both were right.

Fermentation and putrefaction were due to the activities of living cells, but the mechanism whereby the cells brought this about, a point not considered by Pasteur, was that these cells contained the catalyst so fervently believed in by the Berzelius-Liebig school. These catalysts or enzymes were to be the subject of much intense study during the ensuing years and Buchner's experiment was to be repeated in a variety of ingenious ways.

ENZYMES AND ENZYME ACTION

The chemical reactions which occur in the bacterial cell and are responsible for the breakdown of food supplies, the synthesis of new cells and the formation of by-products are catalysed by enzymes.

Enzymes are highly efficient catalysts enabling the anabolic and catabolic reactions to proceed smoothly and at low temperatures. All enzymes so far isolated are proteins and may in addition contain a non-protein fraction called a coenzyme, cofactor, or prosthetic group.

A feature of the catalytic activity of enzymes is their high specificity towards a particular reaction. The compound involved in the reaction is named the substrate for that enzyme.

The specificity of enzyme action is considered to reside in the unique shape that a protein surface can assume; an enzyme protein is able to adsorb its substrate and thereby activate it so that its decomposition or union with another molecule is spontaneously affected at low temperatures, such as the growth temperature of a particular bacterial species.

Coenzymes (prosthetic groups). Enzyme preparations were frequently dialysed during purification and in certain cases this process was found to yield an inactive preparation. It was found

TABLE 5

IMPORTANT COENZYMES

Name	Other names	Function
Nicotinamide-adenine dinucleotide (NAD)	Diphosphopyridine-nucleotide Coenzyme I DPN	Hydrogen transfer
Nicotinamide-adenine dinucleotide phosphate (NADP)	Triphosphopyridine-nucleotide Coenzyme II TPN	Hydrogen transfer
Flavin-mononucleotide (FMN)		Hydrogen transfer
Flavin-adenine dinucleotide (FAD)		Hydrogen transfer
Iron-porphyrin of cytochromes		Electron transfer
Ferredoxin		Electron transfer
Ubiquinone	Coenzyme Q	Electron transfer
Coenzyme A (Co A)		Acyl group transfer
Adenosine 5′-di- and tri-phosphates	ADP, ATP	Phosphate transfer
Pyridoxal-5′-phosphate		Amino group transfer
Pyrophosphothiamine	Cocarboxylase Thiamine pyrophosphate Aneurine pyrophosphate	Decarboxylation of 1-oxo acids Formation of acetoin

that addition of the dialysate to the enzyme resulted in reactivation and this led in turn to the isolation of substances which were found to have much lower molecular weights than the proteins they reactivated. Variation in the degree of binding of the dialysable fraction with the protein led to the use of the term prosthetic group for the more firmly bound dialysable fraction and coenzyme for the less firmly bound. The term coenzyme will be retained here irrespective of the degree of binding. Furthermore, some of the coenzymes isolated were found to be common to several enzyme proteins which catalysed a common reaction, for example the removal of a hydrogen atom. The protein differed for each substrate but the common coenzyme was responsible for the carrying of the hydrogen atom.

Some important coenzymes are listed in Table 5, and illustrated in Figs. 16(*a*)–(*h*), to give an idea of their complex structure. They contain in their molecules growth factors and the role of these substances in the metabolism of the organism will now be apparent.

In some instances metal ions are necessary for enzyme function; Mg^{++}, Mn^{++} and Ca^{++} have been shown to be essential for the functioning of some enzymes. The more important factors affecting enzyme action are discussed below.

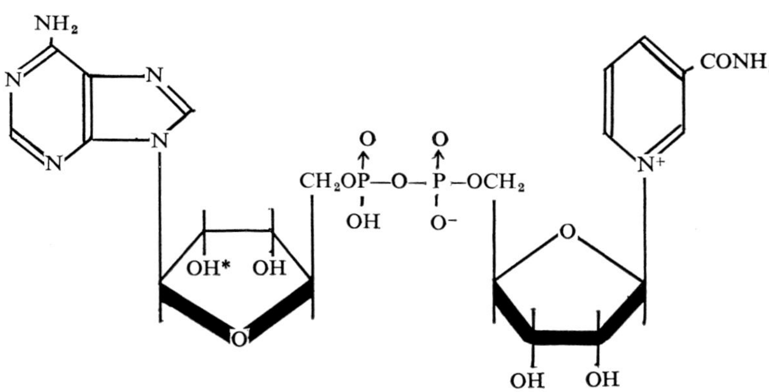

Nicotinamide-adenine dinucleotide (NAD)
(Coenzyme I; diphosphopyridine nucleotide, DPN)

Nicotinamide-adenine-dinucleotide phosphate (NADP)
(Coenzyme II; triphosphopyridine nucleotide, TPN)

The structure of NADP is as for NAD but the —OH group marked with an asterisk is esterified by orthophosphoric acid

FIG. 16(*a*)

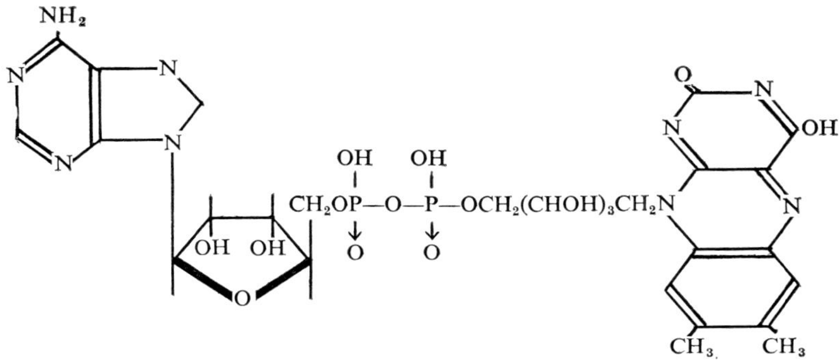

Flavin-adenine dinucleotide
(FAD)

FIG. 16(*b*)

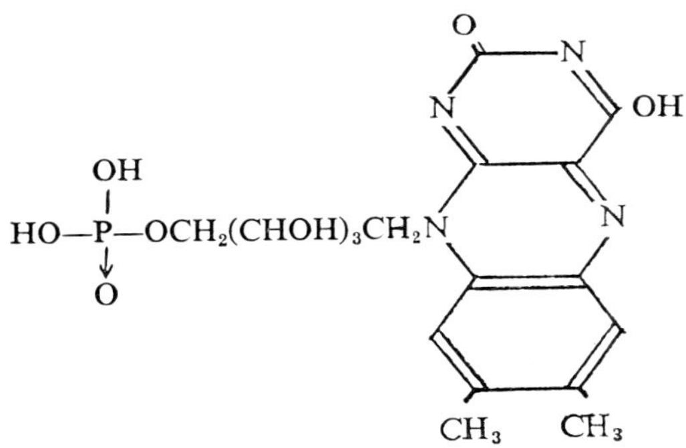

Flavin mononucleotide
(FMN)

FIG. 16(*c*)

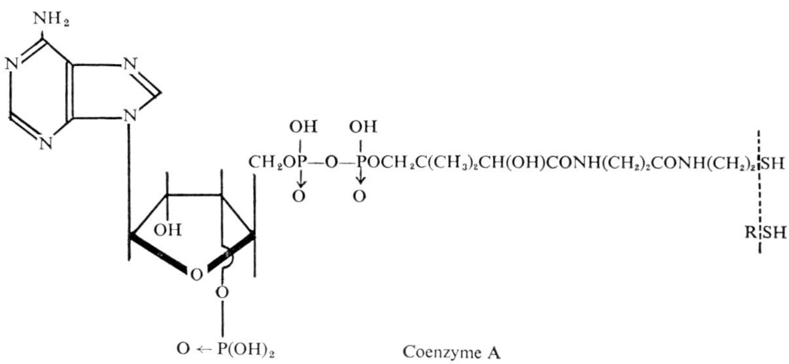

Coenzyme A

$RSCOCH_3$	Acetyl coenzyme A
$RSCOCH_2CH_2COOH$	Succinyl coenzyme A
$RSCOCH_2CH_2CH_3$	Butyryl coenzyme A

FIG. 16(*d*)

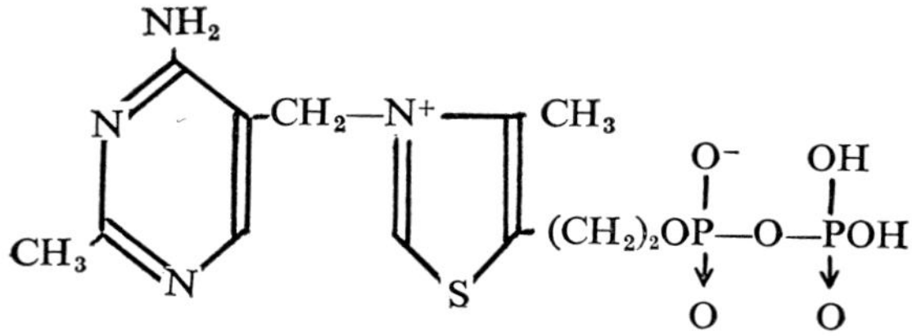

Pyrophosphothiamine
(Cocarboxylase)

FIG. 16(*e*)

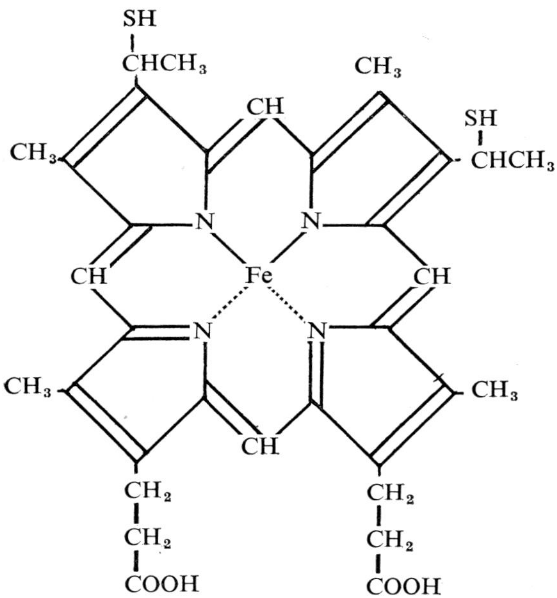

Iron porphyrin of cytochrome *c*

FIG. 16(*f*)

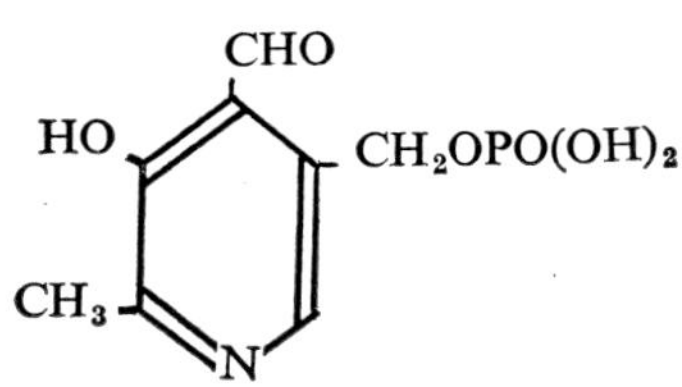

Pyridoxal-5′-phosphate

FIG. 16(*g*)

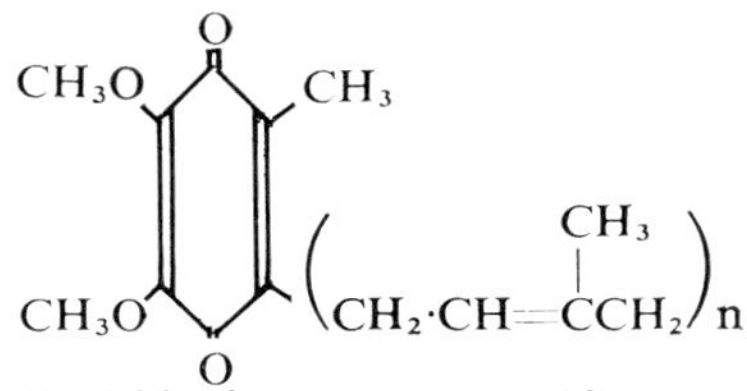

Coenzyme Q. Ubiquinone n = 6 − 10

FIG. 16(*h*)

The Effect of pH

If the velocity of an enzyme reaction is plotted against the pH of the environment in which the reaction is taking place, a typical bell-shaped curve is produced (compare the curve for the effect of pH on bacterial growth). It should be remembered that a change in pH as well as directly affecting the enzyme protein may also affect the coenzyme, if such is implicated in the reaction, and the substrate if the latter is ionised. The experimental curve will represent an integration of all the effects.

The Effect of Substrate Concentration

A plot of velocity of enzyme action against substrate concentration is shown in Fig. 17(*a*). It can be seen that there is a steady increase in velocity up to a concentration represented by $V_{\max}$ thereafter no increase in velocity takes place. It is clear that there is a limit to the amount of substrate that a given amount of enzyme can deal with and an indication of the amount of any particular enzyme may be calculated from this type of experiment. The substrate concentration which produces half-maximum velocity is called the Michaelis constant of the enzyme, K_m.

A curve of the type shown in Fig. 17(*a*) may be transformed to a straight line by plotting $\frac{1}{V}$ against $\frac{1}{S}$, the double reciprocal plot. If the graph is extended backwards, it cuts the x axis at a point equal to $-\frac{1}{K_m}$; from this K_m may be calculated; the minus sign

$$E \quad + \quad S \quad \overset{K_s}{\rightleftharpoons} \quad ES \quad \rightarrow \quad P \quad + \quad E$$

Enzyme	Substrate	Enzyme substrate complex	Product of reaction	Enzyme

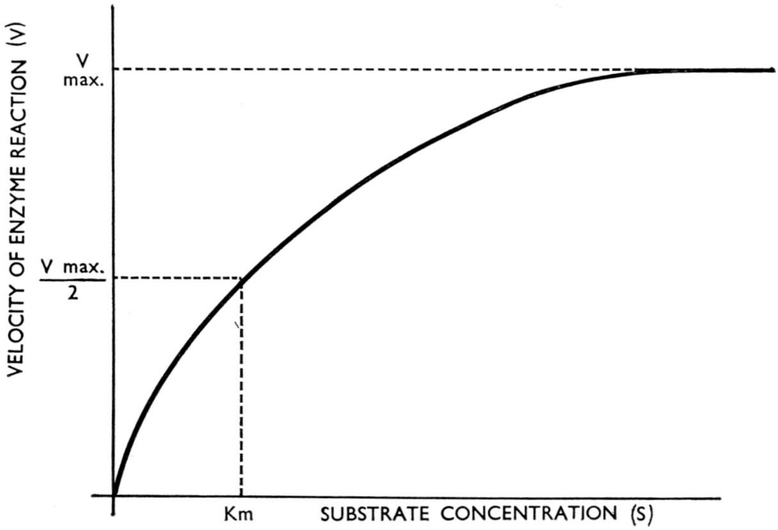

(*a*)

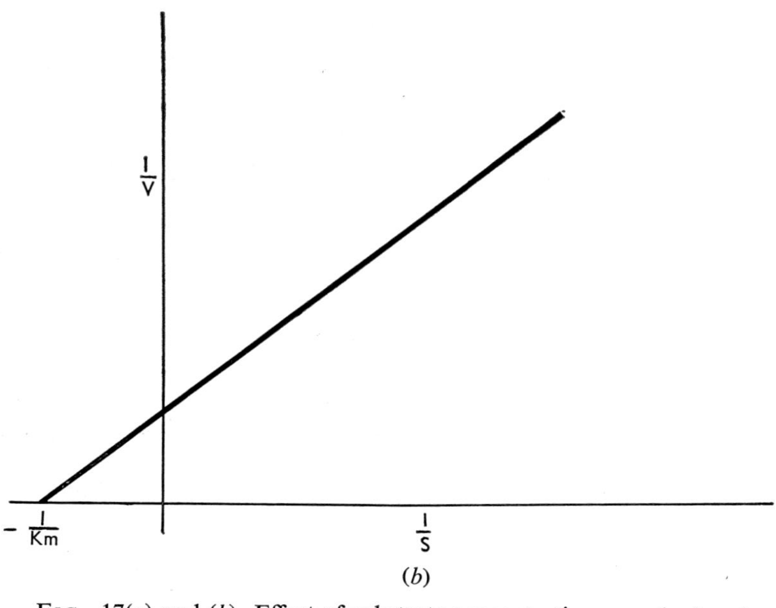

(*b*)

FIG. 17(*a*) and (*b*). Effect of substrate concentration on velocity of enzyme action

has no significance in the value, only identifying the co-ordinate (Fig. 17(*b*)).

It has been shown that during the enzymic decomposition of a compound, the enzyme reacts with the substrate to form an enzyme-substrate complex, and the true rate of reaction is controlled by the formation and subsequent dissociation of this complex. The dissociation constant of the enzyme substrate complex is called the substrate constant, K_s, and is a constant for a particular enzyme.

If a state of equilibrium is maintained between concentration of the enzyme, the substrate and the enzyme substrate complex, then $K_s = K_m$.

The Effect of Temperature

As with the effect of temperature on bacterial growth, a bell-shaped curve can be obtained when enzymic activity is plotted against temperature. The curve again can be thought of as consisting of two parts, one in which the rate of the reaction increases with rise in temperature, following the expected effect of temperature on reaction rate, and a second where, at higher temperatures, the enzyme suffers progressive destruction and the rate of the reaction slows down. Finally, a temperature is reached which inactivates the enzyme completely and no reaction occurs.

Values for the temperature optimum obtained experimentally may vary considerably, depending on the actual time during which the enzyme is allowed to act, but although temperature optima may have little fundamental physico-chemical significance they are of practical value to the technologist.

The Action of Inhibitors

Many antibacterial agents act by their interference with the activity of bacterial enzymes. A useful classification of enzyme inhibitors may be made according to whether they inhibit an enzyme competitively or non-competitively. Competitive inhibition is said to exist when competition can be demonstrated between the inhibitor and the substrate. The degree of inhibition is a function of the ratio of the substrate concentration to the inhibitor concentration. By contrast when the inhibition is non-competitive, the concentration of substrate has no effect on the degree of inhibition, which is dependent on the inhibitor concentration only. It is possible to determine whether an enzyme inhibitor is acting competitively or non-competitively by studying its action at varying substrate concentrations. The effect of substrate concentration on enzyme activity and the derivation of the Michaelis constant has already been mentioned. If the experiment shown in Fig. 17(*a*) is repeated in the presence of an inhibitor, the form of the curve obtained differs according to the nature of the inhibition process (Fig. 18). In non-competitive inhibition, the Michaelis constant remains unaltered but the value of the maximum velocity is de-

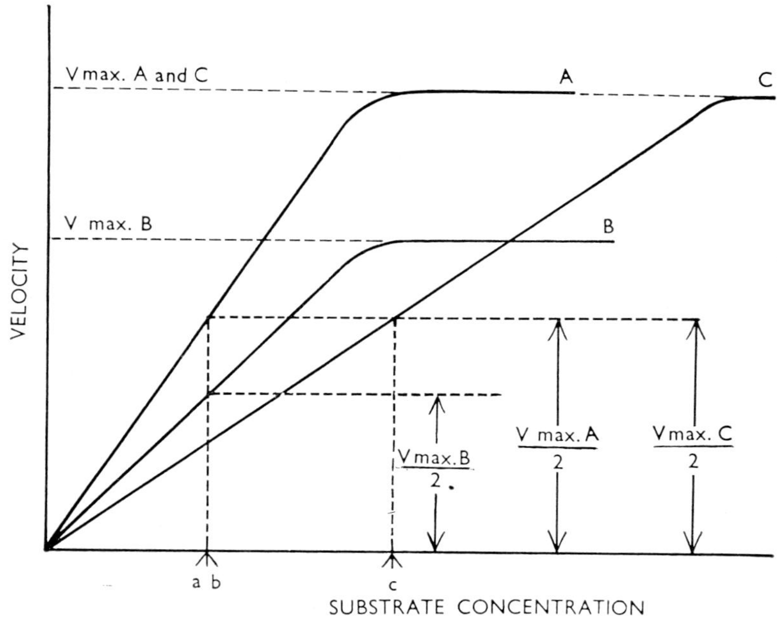

FIG. 18. Effect of inhibitors on enzyme catalysed reactions. A—Uninhibited reaction. B—Non-competitively inhibited reaction. C—Competitively inhibited reaction. a, b, c—substrate concentrations for half maximum velocity ($= K_m$) for reactions A, B and C

creased, while in competitive inhibition the maximum velocity equals that of the uninhibited reaction, but the value of the Michaelis constant is changed. Put in another way, a non-competitive inhibitor appears to act by damaging the actual enzyme itself, thus the velocity of reaction, a function of enzyme concentration, is changed. In competitive inhibition, the effect of the inhibitor is to reduce the effective substrate concentration, thereby altering the Michaelis constant.

In practice the effect of inhibitors on enzyme kinetics is usually studied after transforming the plots shown in Fig. 14 to straight lines.

Constitutive and Induced Enzymes

Quantitative studies of the amount of enzymic material per unit quantity of cellular material have shown a variety of different values and furthermore these values have been shown to vary according to the conditions of growth of the cells concerned. For example the presence of a specific enzyme substrate during growth of bacterial cells may result in a cell crop with a many hundredfold increase in the subsequent ability to decompose that substrate when compared with cells grown in the absence of such a substrate.

A consideration of these facts has led to the recognition of two types of enzyme within the cell and called constitutive enzymes and induced enzymes. Constitutive enzymes are formed in the cell whether the specific substrate is present or not. Induced enzymes are formed in greatly increased amounts when the organisms are grown in the presence of the substrate. The enzymic level may be so low in cells not grown in the presence of substrate as to warrant the assumption that it is zero. The factors affecting induced enzyme formation in bacteria have been the subject of intensive study and much is now known of the process. The change is non-genetic and the enzyme formation is induced by the substrate. Even more interesting is the fact that examples are well substantiated where substances closely related to the substrate can act as an inducer although not being capable of being metabolised.

BACTERIAL METABOLISM

Metabolic processes may be divided into two categories, anabolic or synthetic and catabolic or degradative. Overall, during metabolism, energy must be derived from catabolic processes to drive anabolic processes.

The study of bacterial metabolism involves the measurement of the rate of disappearance of substances, the detection and estimation of reaction products, the measurement of the rate of use of oxygen during respiration and many specialised techniques such as the spectroscopic determination of respiratory pigments, to mention only one example. Much use is made of washed suspensions of bacteria (prepared by suspending bacteria in water or in solutions containing various combinations of ions) in metabolic studies, it being argued that during an experiment with washed suspensions, no growth occurs and chemical changes uncomplicated by changes due to growth may thus be studied. A further experimental stage is to attempt to prepare from the bacterial cells the enzymes responsible for the changes being studied, and to determine the nature of any coenzyme or metal ions that may be implicated in the enzyme-catalysed reaction.

Oxidative Processes in the Bacterial Cell

Bacteria are able to oxidise a variety of substances by a series of different methods often involving complex cycles. The oxidative breakdown of these compounds usually yields energy to the cell for growth and other synthetic processes. Atmospheric oxygen may be the oxidising agent but in bacteria growing anaerobically it is obvious that an oxidising agent other than oxygen is being used.

Oxidative processes in the cell usually involve hydrogen transport and the type reactions that occur for the oxidation of a hypothetical compound SH_2 are

$$2SH_2 + O_2 = 2S + 2H_2O \quad \text{aerobic}$$
$$SH_2 + X = S + H_2X \quad \text{anaerobic}$$

In the first reaction the oxidising agent is oxygen, which is reduced to water, this is an aerobic reaction; in the second some other substance, X (not oxygen) acts as the oxidising agent and is reduced to H_2X; this is an anaerobic reaction.

A variety of terms is used in describing these reactions and it is necessary to define them.

Respiration. This term will be used to denote the uptake of oxygen by bacteria, and thus is used in the same sense as in mammalian and plant physiology.

Fermentation. This is an anaerobic process and refers to the anaerobic breakdown of chemical substances.

Dehydrogenation. This term is self-explanatory; it means the removal of hydrogen from a substance and implies that that substance undergoes oxidation as a result. The hydrogen may react finally with gaseous oxygen as in respiration or with another substance as in fermentation. Hydrogen may be removed in one of two ways, either as H, the hydrogen atom, or $H^+ + e$, a hydrogen ion plus one electron.

Hydrogen or electron donor. Any substance which, during metabolism, yields hydrogen either as H or as $H^+ + e$.

Hydrogen electron acceptor. A substance which reacts with hydrogen (H or $H^+ + e$). This, in respiration, is oxygen, in fermentation some other hydrogen acceptor.

Oxidation. This term may cover the following reactions: the uptake of oxygen, loss of hydrogen and/or the loss of electrons.

The term respiration is used in some texts to cover both aerobic and anaerobic energy-yielding reactions; it is felt that as stress has always been laid on the difference between aerobic and anaerobic bacteria, the terms to distinguish their metabolism should be retained. In the industrial field, the highly aerobic processes involved in antibiotic production are often referred to as fermentations and vessels with forced aeration as fermenters. This terminology is a legacy from the brewing industry.

THE PATTERN OF OXIDO-REDUCTION IN BACTERIA

The principles of oxido-reduction, the nature of enzyme action of co-enzymes and of redox potentials have already been dealt with. In this section, some examples of both aerobic and anaerobic pathways for linked oxido-reductions in bacteria will be considered. A summary is presented in Fig. 19, p. 89.

The Complete Aerobic Chain

The most complex path is that from a substrate, S_fH_2, to molecular oxygen involving all the known components of the aerobic oxido-reduction chain and this path and its components will be considered first.

In the overall process hydrogen is removed from a substrate and travels as H or $H^+ + e$ to molecular oxygen reducing the latter to water. It will be recalled that each component of the redox chain has a characteristic redox potential (E_0') and the direction of movement in the chain is always in the direction low to high redox potential. The components are:

	$E_0'V$ in the region of
1. Nicotinamide adenine dinucleotide and phosphate	−0·32
2. Flavine mononucleotide; Flavine adenine dinucleotide	−0·06
3. Coenzyme Q. Ubiquinone	+0·10
4. Non-haem iron	?
5. The cytochrome system	from +0·002 to +0·29

Components 1–3 carry hydrogen atoms as $H^+ + e$; the cytochromes carry electrons only; the hydrogen ion with the electron

is finally united under the enzymic activity of cytochrome oxidase to reduce oxygen to water.

Nicotinamide adenine dinucleotide, NAD, and Nicotinamide adenine dinucleotide phosphate, NADP. These two components, Fig 16 (a), were originally known as coenzymes I and II as they are able to function with many hydrogen removing (dehydrogenase) enzymes each specific for its substrate.

They have redox potentials, E_0', of the order $-0{\cdot}32$ *V* and are in aerobic and some anaerobic reactions, the first in the chain. Their alternating oxido-reduction may be represented thus, $NAD^+ \leftrightharpoons NADH$ and $NADP^+ \leftrightharpoons NADPH$.

Flavine mononucleotide, FMN, and Flavine adenine dinucleotide FAD. These two co-enzymes are illustrated in Fig. 16 (a) and (b), and are derivatives of riboflavine (vitamin B_2). The bright yellow colour conferred by the riboflavine when linked with an appropriate enzyme protein caused these to be called yellow enzymes when they were first isolated. The enzymes are now called flavoproteins. Their alternating oxido-reduction may be represented thus, $FMN \leftrightharpoons FMNH_2$ and $FAD \leftrightharpoons FADH_2$.

Co-enzymes Q, CoQ, Ubiquinones. This group was discovered in 1957 and the structure of the co-enzyme illustrated in Fig. 16 (h). They link the flavoproteins with the cytochrome system and may be represented thus $CoQ \leftrightharpoons CoQH_2$.

Non-haem iron. In many experiments in the purification of the components of the redox or respiratory chain, iron compounds differing from the cytochromes (see below) were isolated. This so-called non-haem iron has been thought to participate in the chain but work is still proceeding on its more precise function. It is not included in Fig. 19.

The cytochromes. These, the last members of the chain and which react with oxygen itself, are a group of conjugated proteins, known as haemoproteins. The non-protein part of the molecule is related to haem and is an iron porphyrin (Fig. 16(*f*)). These pigments occur widely in nature and act as reversible oxidation-reduction systems by virtue of a change in the valency of the iron atom, $Fe^{++} \leftrightharpoons Fe^{+++}$.

Spectroscopy, with instruments from the simplest direct vision spectroscope to self-recording spectrophotometers capable of measuring spectral changes in times of the order of milliseconds, has played a large part in the study of the cytochrome pigments. Sixteen different cytochromes have been described at the time of writing and are differentiated into cytochromes *a*, *b* or *c* by spectral and chemical characteristics.

Cytochrome a group. Absorption bands between 590 and 635 nm when reduced, that is with the iron in the molecule in the ferrous state; cytochromes *a* to a_4.

Cytochrome b group. Bands at 560 and 530 nm when reduced members of this group are split into haem and protein by treatment with alkalies; cytochromes *b* to b_7.

Cytochrome c group. Members of this group are more stable towards heat, acid and alkali than members of group *a* or *b*; cytochromes *c* to c_5.

The cytochrome carriers differ from haemoglobin in that they are not spontaneously reoxidised by the oxygen in the air but require an enzyme called cytochrome oxidase for this step; cytochrome oxidase is related to the cytochromes in being an iron porphyrin conjugated with a protein. In the nomenclature given above, cytochrome oxidase is a mixture of cytochrome *a* and a_3. In some bacteria the terminal oxidase has been found to differ from the *a* to a_3 type and has been designated cytochrome *o*. It is the only cytochrome oxidase in some staphylococci. Individual bacterial species have been found to differ in their complement of the cytochrome carrier pigments, some possessing one, some two, some all three. The reaction sequence in the cytochromes is as follows. The figures are the E_0 values, at pH 7 and 30°, in volts.

→ *b* group if present →	*c* group if present →	*a* group if present →	cytochrome oxidase →	oxygen
−0·062	+0·25	+0·29		

——————— progressive reduction ———————→

The overall reaction in the complete aerobic chain is

$$S_fH_2 \rightarrow S_f/\tfrac{1}{2}O_2 \rightarrow H_2O. \text{ (Path 6, Fig. 19).}$$

A component, ferredoxin which occurs in anaerobic bacteria will be referred to below.

Anaerobic Oxido-reductions Involving the Ferredoxins

Anaerobic bacteria, which can operate at low E_0' values (page 56) were shown in 1962 to contain another group of non-haem iron proteins named ferredoxins. The ferredoxins have E_0' values of the order of −0·411 *V* and take part in an oxido-reduction process, via NAD, between two substrates one of which is oxidised while one is reduced. This is path 1 of Fig. 19 and the overall reaction is $S_aH_2 \rightarrow S_a/S_b \rightarrow S_bH_2$.

Oxido-reductions via NAD and NADP

These coenzymes again when linked with their specific dehydrogenases may take place in an anaerobic hydrogen transfer thus $S_cH_2 \rightarrow S_c/S_d \rightarrow S_dH_2$ (Path 2 of Fig. 19). Examples of this reaction are found in the anaerobic decomposition of carbohydrates by micro-organisms, see page 19, and the possession of specific enzymes, coupled with NAD or NADP, account for the varying patterns found which in turn accounts for the use of sugars, alcohols and glycosides in diagnostic tests, in which the media is designed to show the production of acid and sometimes gaseous products from the original substrate in the medium, p. 125.

Oxido-reductions via FAD or FMN

In these reactions water is reduced to hydrogen peroxide which, if catalase is present, is, decomposed to water and oxygen, $S_eH_2 \rightarrow S_e/O_2 \rightarrow H_2O_2$, path 3 of Fig. 19.

The amino acid oxidases are examples of this type of reaction wherein an amino acid is oxidised to an oxo acid.

Another example is the glucose oxidase found in many moulds. It catalyses the oxidation of glucose to gluconic acid with production of hydrogen peroxide. This metabolic sequence gave rise to the idea that certain moulds that carried it out were producing an antibiotic. Hydrogen peroxide accumulation, due to absence of catalase in the organism, was the cause of the antibacterial effect.

In addition to the oxidases containing a derivative of riboflavine, there is a group of enzymes called oxygenases which consist of a protein linked with a metal such as iron or copper. Some enzymes capable of oxidising phenols to quinones belong to this class. Oxidase and oxygenases reactions yield little, if any, energy to the cell.

In some bacterial species e.g. the pneumococci and some staphylococci, cytochromes are absent and oxygen is reduced via auto-oxidisable flavoproteins.

Succinic Acid Reduction

Succinic acid, a component of the Krebs cycle, enters the oxidation/reduction chain at the flavoprotein level. Thereafter the normal reaction proceeds via CoQ, and the cytochromes resulting in the reduction of molecular oxygen to water. (Path 4, Fig. 19)

$$\text{succinic acid} \rightarrow \text{fumaric acid}/\tfrac{1}{2}O_2 \rightarrow H_2O.$$

Oxidation of formic acid

Formic acid enters the chain at the cytochrome *b* level and is oxidised to carbon dioxide via the cytochromes and cytochrome oxidase while oxygen is reduced to water i.e.

$$H{\cdot}COOH \rightarrow CO_2 \rightarrow H_2O.$$

(Path 5 of Fig. 19).

These examples illustrate the complexity of biological oxidation/reduction patterns encountered in bacteria.

Facultative Organisms and the Pasteur Effect

It is interesting to speculate upon the behaviour of a facultative organism when growing with adequate oxygen supply. Pasteur noted that in this group of organisms, growth is far more rapid under aerobic conditions, as compared with anaerobic conditions, but despite this the utilisation of carbohydrate under anaerobic conditions is greater. This phenomenon is known as the Pasteur effect and suggests that aerobic metabolism of carbohydrates is a more efficient process than the anaerobic metabolism in organisms which can carry out both types of metabolic reactions.

Energy in the Living Cell

Mention has been made during discussion of oxidation-reduction processes of energy being derived for anabolic reactions by the breakdown or catabolism of a variety of inorganic and organic compounds. Energy is stored and released in living organisms mainly through the agency of esters of pyrophosphoric acid, but also as thiol esters such as derivatives of coenzyme A.

One of the most important esters is a derivative of adenosine; it was found that phosphate esters of this compound occur in cellular material and that the energy released on hydrolysis of adenosine triphosphate (ATP) (Fig. 20), to adenosine diphosphate (ADP) or to hydrolyse ADP to adenosine monophosphate (AMP) was more than three times that released on hydrolysis of adenosine monophosphate to adenosine (A). In the terms of thermodynamics the free energy of hydrolysis of ATP to ADP and ADP to AMP is more than three times the free energy of hydrolysis of AMP to A.

Key to FIG. 19

1. $S_aH_2/Sb \rightarrow S_bH_2$ Anaerobic oxido-reduction between substrates S_a and S_b via ferredoxin (Fd) and NAD. Found in strict anaerobes.
2. $S_cH_2 \rightarrow S_c/S_d \rightarrow S_dH_2$ Anaerobic oxido-reduction between substrates S_c amd S_d via NAD or NADP.
3. $S_eH_2 \rightarrow S_e/O_2 \rightarrow H_2O_2$ Oxidation of S_eH_2 to S_e via FAD or FMN with reduction of oxygen to hydrogen peroxide. Hydrogen peroxide converted to water by catalase.
4. Succinic acid $\rightarrow$ fumaric acid/$\frac{1}{2}O_2 \rightarrow H_2O$ Oxidation of succinic acid to fumaric acid via a flavoprotein, coenzyme Q and the cytochrome system with reduction of oxygen to water.
5. Formic acid $\rightarrow CO_2/\frac{1}{2}O_2 \rightarrow H_2O$ Oxidation of formic acid to CO_2 via the cytochrome system with reduction of oxygen to water.
6. $S_fH_2 \rightarrow Sf/\frac{1}{2}O_2 \rightarrow H_2O$ Oxidation of S_fH_2 to S_f via the complete oxidation reduction chain with reduction of oxygen to water.

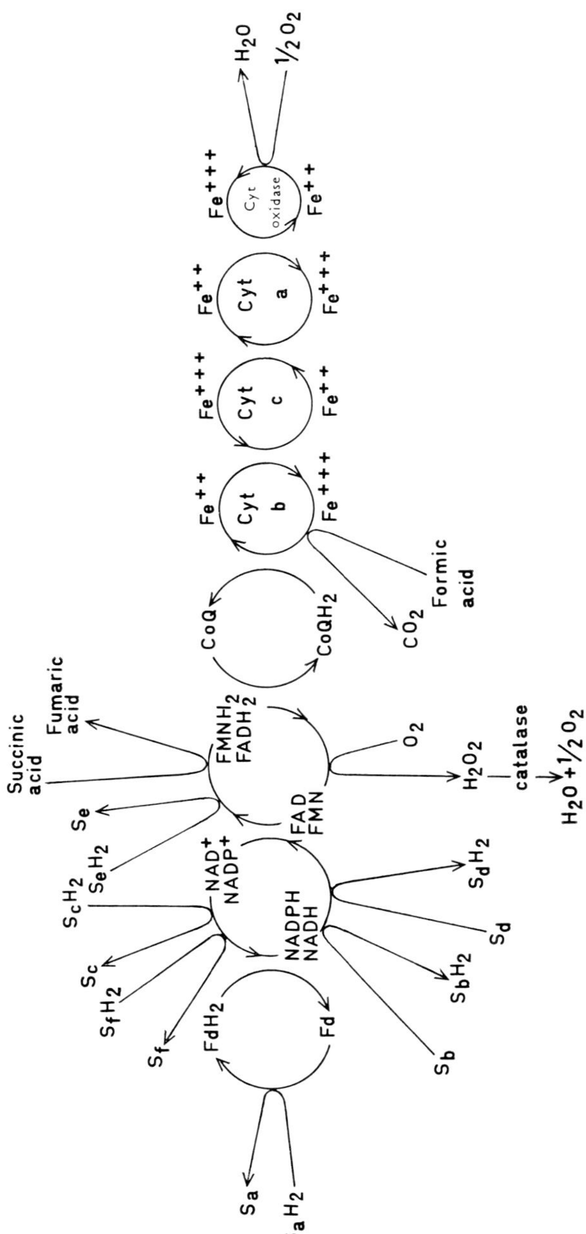

FIG. 19. Patterns of aerobic and anaerobic oxidation and reduction in bacteria. (With the exception of cytochrome oxidase, the related enzymes have not been shown.)

Lipman suggested in 1941 that high-energy phosphate bonds are a means of energy transfer and storage in the cell.

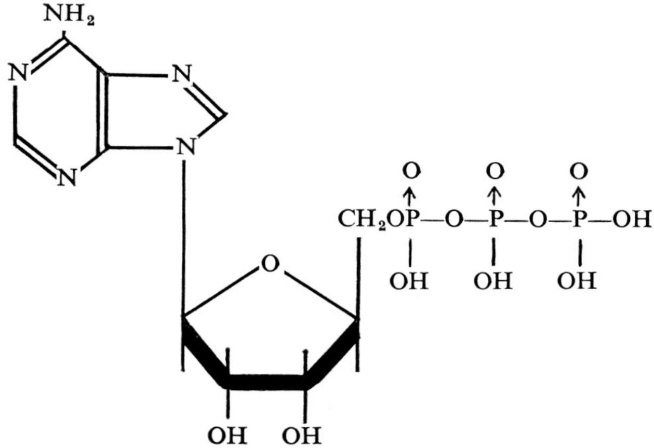

Adenosine triphosphate
(ATP)

FIG. 20

Thus an energy-yielding catabolic reaction may result in the esterification of ADP to yield ATP, with an uptake of inorganic phosphate Pi. While later an energy-requiring anabolic reaction may result in the hydrolysis of ATP to yield ADP again; the phosphate bond energy, as it is sometimes termed, being made available for the anabolic process.

The two reactions, esterification or phosphorylation and hydrolysis, are carried out by specific enzymes and the catabolic and anabolic reactions occur simultaneously, the ATP acting as a carrier of energy much as the coenzymes and cytochromes carried hydrogen and electrons.

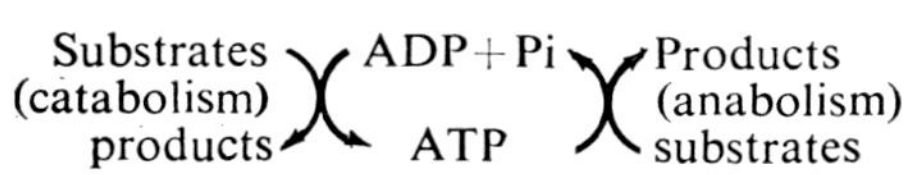

Some light is shed on the remarks made under the heading of the Pasteur effect by the fact that in a facultative organism, the anaerobic decomposition of 1 mole of glucose yielding ethanol or lactic acid results in the release of energy sufficient only to generate 'high-energy' phosphate bonds. If 1 mole of glucose is oxidised aerobically to completion, via the complete respiratory dream and with no accumulating by-product, 30 high-energy phosphate bonds are synthesised. Thus the aerobic decomposition of glucose results in a much greater use of the chemical energy of the glucose molecule. This combination of oxidation with phosphorylation is known as oxidative phosphorylation. In addition to oxidative phosphorylation, ATP may be generated at the substrate level, the so-called substrate level phosphorylation. In this process, if substrate A is converted enzymatically to substrate B with sufficient energy output then, in the presence of inorganic phosphate, ATP may be generated from ADP. An example is the conversion of glyceraldehyde-3-phosphate to 3-phosphoglyceric acid in the Embden–Meyerhof pathway for anaerobic glucose catabolism described immediately below.

THE ANAEROBIC CATABOLISM OF GLUCOSE

The complex pattern involved in the breakdown of glucose by micro-organisms received early attention because of the great interest of scientists in the mechanism of the alcoholic fermentation. A similar mechanism for anaerobic glycolysis is involved during muscular activity in higher organisms, the final product here being, however, lactic acid and not ethanol.

The Embden–Meyerhof glycolysis cycle for the conversion of glucose to ethanol is shown in part in Fig. 21. It is called a cycle because, as will be seen, there are within the sequence of reactions which lead to the production of ethanol two processes, one of which involves the generation and breakdown of adenosine-triphosphate, and the other the generation and breakdown of reduced nicotinamide-adenine dinucleotide (NAD).

The first stage in the glycolysis cycle involves the phosphorylation and isomerisation of glucose, which in natural fermentations may arise from the preliminary breakdown of starches or sucrose.

The glucose phosphorylation step is catalysed by the enzyme hexokinase, the phosphate radical is supplied by adenosine-triphosphate, which is converted to adenosine-diphosphate in the process. The result is glucose-6-phosphate. In addition to the enzyme and ATP, magnesium ions are essential for this reaction. Any reagent, such as sodium fluoride, which removes magnesium ions by precipitation, will inhibit the glycolysis cycle.

Glucose-6-phosphate, under the influence of an enzyme phosphohexose isomerase, undergoes an isomeric change to yield a phosphate ester of fructose, fructose-6-phosphate. At this stage a second phosphorylation occurs, the phosphate radical again being supplied by ATP; the enzyme is called phosphofructokinase and

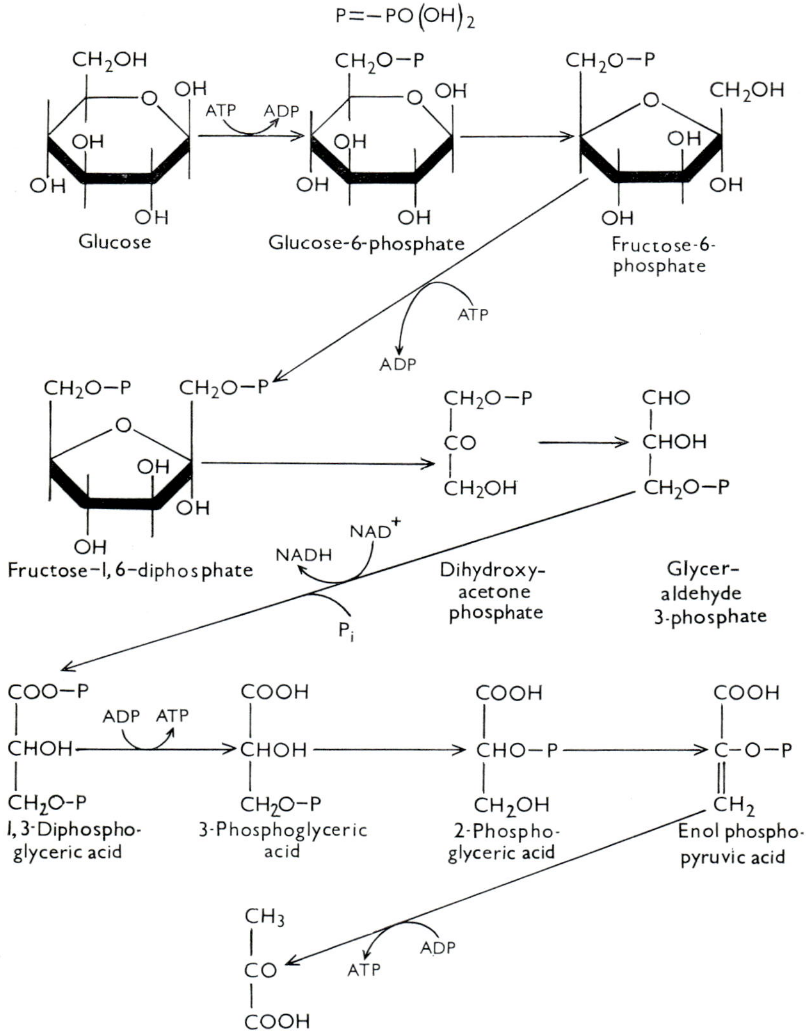

FIG. 21. The Embden–Meyerhof pathway for the degradation of glucose to pyruvic acid.

also requires magnesium ions for its function; the reaction is also inhibited by reagents which inactivate magnesium. The result of this reaction is the production of fructose-1,6-diphosphate.

The second stage consists of the splitting of the 6-carbon sugar ester to yield two 3-carbon compounds, again esters of phosphoric acid. An enzyme called aldolase converts fructose-1,6-diphosphate to an equilibrium mixture of glyceraldehyde-3-phosphate and dihydroxyacetone phosphate; the latter is converted by the enzyme triose phosphate isomerase to more glyceraldehyde-3-phosphate. The aldolase, triose phosphate isomerase mixture is sometimes referred to as zymohexase. The third stage involves an oxidative step catalysed by NAD. Glyceraldehyde-3-phosphate is oxidised

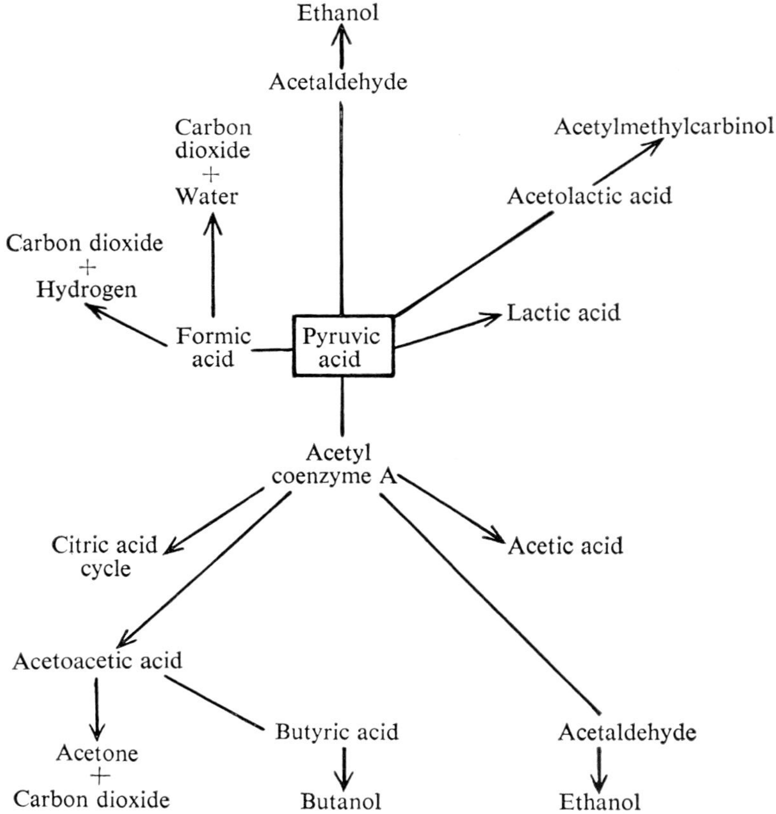

FIG. 22. Products of the decomposition of pyruvic acid by micro-organisms

to 3-phospho-glyceric acid, which also undergoes phosphorylation so that the final product is 1,3-diphosphoglyceric acid. Inorganic phosphate (Pi) serves as a source of the phosphate radical in this reaction. The enzyme responsible for the oxidative step is called glyceraldehyde-phosphate dehydrogenase. The reduced NAD is available as a reducing agent for further metabolic changes that occur later in the sequence of reactions leading to ethanol.

The next reaction involves the hydrolysis of 1,3-diphosphoglyceric acid to 3-phosphoglyceric acid, ATP being resynthesised. Phosphoglyceromutase now converts 3-phosphoglyceric acid to 2-phosphoglyceric acid and enolase converts 2-phosphoglyceric acid to enolphosphopyruvic acid, magnesium ions also being necessary for this reaction. This is hydrolysed by an appropriate phosphatase to pyruvic acid with the regeneration of a molecule of ATP from ADP.

Diversities in metabolic pathways of anaerobic glucose breakdown are associated with diversities in the further breakdown of pyruvic acid (Fig. 22). The various fermentation paths shown are sometimes named by the predominant end-product. Alternatively the fermentation may be named after the organism or group of organisms in which it is found.

Conversion of Pyruvic Acid to Ethanol by Yeast

This sequence of reactions is a continuation of those considered above. Pyruvic acid is decarboxylated to yield acetaldehyde by the enzyme pyruvic acid decarboxylase. This enzyme has as its coenzyme, aneurine diphosphate. Finally, acetaldehyde is reduced to ethanol. The enzyme catalysing this reaction is called alcohol dehydrogenase although the dehydrogenation, that is the oxidation of alcohol, is the reverse of the reaction occurring. The coenzyme of yeast alcohol dehydrogenase is NAD and the reduced NAD formed during the oxidation of glyceraldehyde-3-phosphate to 1,3-diphosphoglyceric acid is now reconverted to NAD as the acetaldehyde is reduced to ethanol.

Formation of Ethanol by *Escherichia coli*

Esch. coli produces very small quantities of ethanol but the mechanism of production differs from that found in yeast. Glucose is converted to pyruvic acid according to the Embden–Meyerhof scheme as for yeast, thereafter pyruvic acid is converted to formic acid and acetyl coenzyme A. Coenzyme A (Fig. 16(*d*)) acts as a carrier of acyl groups during metabolic reactions. Two enzymes, both having NAD as their coenzyme, convert the acetyl coenzyme A first to acetaldehyde then to ethanol.

Lactic Fermentation

Lactic acid is produced from pyruvic acid by the action of lactic dehydrogenase. Organisms producing lactic acid have been divided into two categories. Those producing exclusively lactic acid are called homolactic fermentors, an example is *Streptococcus lactis*; those organisms producing lactic acid and other products are termed heterolactic fermentors, an example is *Lactobacillus brevis*, which in addition to producing lactic acid produces acetic acid, ethanol and glycerol.

Acetone–Butanol Fermentation

Acetone for the manufacture of explosives was prepared during the 1914–18 war by fermentation of carbohydrate waste such as crude molasses or ground maize, using the anaerobic organism *Clostridium acetobutylicum*. A second by-product of this fermentation was butanol, for which at first no use could be found but later butanol found extensive application as a lacquer solvent. With the development of the petrochemical industry, the production of acetone and butanol by fermentation has decreased to the point of obsolescence in the United Kingdom. Fermentation is still used to produce these solvents in India and Australia, however.

The complex reactions now recognised in the formation of butanol and acetone, involving as they do both phosphate esters and coenzyme A derivatives, may be simplified verbally as follows: the sucrose in the raw molasses is converted to glucose and then via the Embden–Meyerhof cycle to pyruvic acid; pyruvic acid is then converted to acetate, acetyl phosphate, carbon dioxide and water. One molecule of acetyl phosphate and one molecule of acetic acid now condense to yield one molecule of acetoacetic acid. Acetoacetic acid is further metabolised according to the simplified scheme given on page 96.

As can be seen, acetone arises from the decarboxylation of acetoacetic acid, carbon dioxide being the other reaction product. Butanol arises by successive reductions; the scheme shown is simplified in that the intermediates react as derivatives of coenzyme A, for example butyryl coenzyme A.

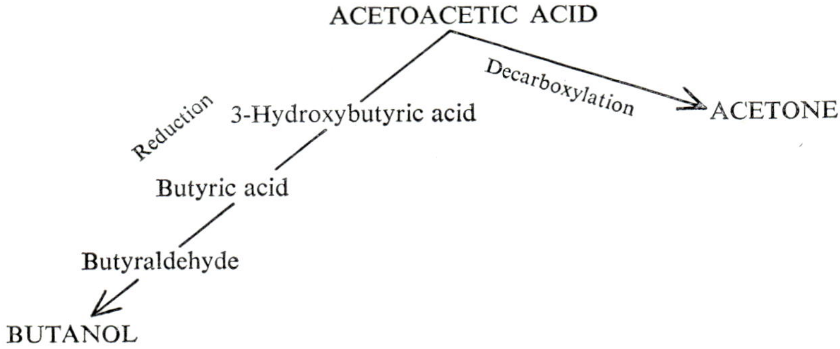

Other products of this fermentation include acetic and butyric acids, ethanol and acetylmethylcarbinol. The required products are obtained from the fermentation mixture by fractional distillation.

Acetylmethylcarbinol Formation

The compound $CH_3{\cdot}CH{\cdot}OH{\cdot}CO{\cdot}CH_3$, butan-2-ol-3-one, acetoin or acetylmethylcarbinol is a product of glucose fermentation in some bacterial species and the ability of certain groups of closely related organisms to form acetoin or not is used as a tool in diagnostic bacteriology.

Glucose is converted to pyruvic acid by the Embden–Meyerhof scheme and acetoin arises from the combination of two molecules of pyruvic acid with elimination of one molecule of carbon dioxide to yield one molecule of acetolactic acid. Two enzymes are involved in this reaction, one effecting the combination of two molecules of pyruvic acid, the other the decarboxylation reaction. Cocarboxylase is required for this reaction sequence. Acetolactic acid now loses a molecule of carbon dioxide to give acetoin, in a reaction catalysed by acetolactic decarboxylase.

$$2CH_3{\cdot}CO{\cdot}COOH \longrightarrow \underset{\text{Acetolactic acid}}{CH_3{\cdot}CO{\cdot}\overset{\overset{\displaystyle OH}{|}}{\underset{\underset{\displaystyle CH_3}{|}}{C}}{\cdot}COOH} + CO_2$$

$$\underset{\text{Acetolactic acid}}{CH_3{\cdot}CO{\cdot}\overset{\overset{\displaystyle OH}{|}}{\underset{\underset{\displaystyle CH_3}{|}}{C}}{\cdot}COOH} \longrightarrow \underset{\text{Acetoin}}{CH_3{\cdot}CO{\cdot}\overset{\overset{\displaystyle OH}{|}}{\underset{\underset{\displaystyle CH_3}{|}}{CH}}} + CO_2$$

The Voges–Proskauer reaction. This reaction, discovered in 1898, detects acetoin. Voges and Proskauer had noticed that if potassium hydroxide solution is added to cultures of certain bacterial species, a red fluorescence is produced after standing for 24 hours. This was traced to the following sequence of events. Acetoin, produced as outlined above, is oxidised under alkaline conditions by atmospheric oxygen to the corresponding diketone, diacetyl, $CH_3{\cdot}CO{\cdot}CO{\cdot}CH_3$. This ketone reacts with compounds containing a guanidine group, for example creatine or guanidine residues in bacteriological culture media made from proteins, to give the red fluorescent compound.

Although a positive Voges–Proskauer reaction is given by members of a number of bacterial genera, *Bacillus*, *Staphylococcus* and *Serratia*, for example, it has found particular application in distinguishing two organisms, *Esch. coli* and *Aerobacter aerogenes*. The practical significance in being able to distinguish unequivocally between these two organisms lies in the fact that if isolated from domestic water supplies, the presence of the former may indicate pollution by sewage and therefore the potential danger of more serious infections arising, for example food poisoning, dysentery or typhoid fever. If the organism is proved to be *Aerobacter aerogenes*, it is likely to have arisen as a chance contaminant, as it is present in dust or soil, and its incidence in the intestine is low. *Esch. coli* does not produce acetoin from glucose, whereas *Aerobacter aerogenes* does.

Formation of Hydrogen

Coliform organisms carry out a series of reactions, in which phosphate esters and derivatives of coenzyme A are implicated and in one of which hydrogen is evolved (Fig. 23). The point of departure from the common path of carbohydrate breakdown is pyruvic acid.

Pyruvic acid reacts with coenzyme A to form acetyl coenzyme A and formic acid. Acetyl coenzyme A, via acetyl phosphate, yields acetic acid. It will be recalled that reduction of acetyl coenzyme A by enzymes in *Esch. coli* gives rise to ethanol (page 94).

The formic acid so formed may be decomposed by an enzyme complex named formic hydrogenylase, giving hydrogen and carbon dioxide, which constitutes the 'gas' that collects in the fermentation tube in carbohydrate fermentation tests.

Formic dehydrogenase in the presence of cytochrome b_1 decomposes formic acid in a different manner to give carbon dioxide and reduced cytochrome b_1. In the presence of oxygen and cytochrome oxidase the final reaction products will be water and carbon dioxide.

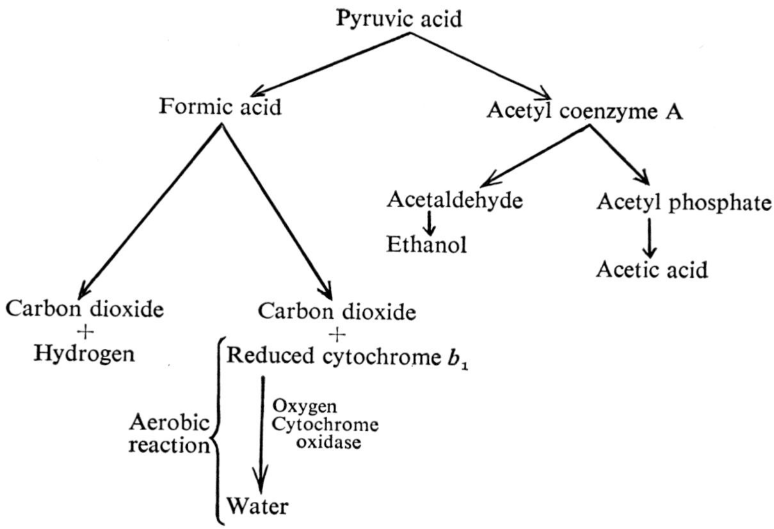

FIG. 23. Decomposition of pyruvic acid by coliform organisms

THE KREBS' CITRIC ACID CYCLE

The reactions so far considered represent various anaerobic routes, for the breakdown of carbohydrate and pyruvic acid was seen to be a point of departure for the differing pathways found in micro-organisms.

There exists in nature a cycle involving nine intermediate compounds and eight enzymes for the *aerobic* oxidation of pyruvic acid to water and carbon dioxide. This cycle was first described by Krebs in 1937 and is known as the Krebs' tricarboxylic or citric acid cycle. It occurs in plants and animals and in many but not all bacteria, capable of growing aerobically.

It is shown diagrammatically in Fig. 24. Pyruvic acid is first converted to acetyl coenzyme A, which, under the influence of an enzyme called the condensing enzyme or citrogenase, combines with oxaloacetic acid to yield citric acid; this reaction initiates the cycle. At this stage one molecule of carbon dioxide is evolved.

The oxaloacetate required for this initial step can arise from a transamination reaction involving glutamic acid, pyruvic acid or oxaloacetic acid, or in *Esch. coli* for example by the dehydrogenation and deamination of glutamic acid. Eventually α-ketoglutaric acid will be regenerated in the Krebs' cycle.

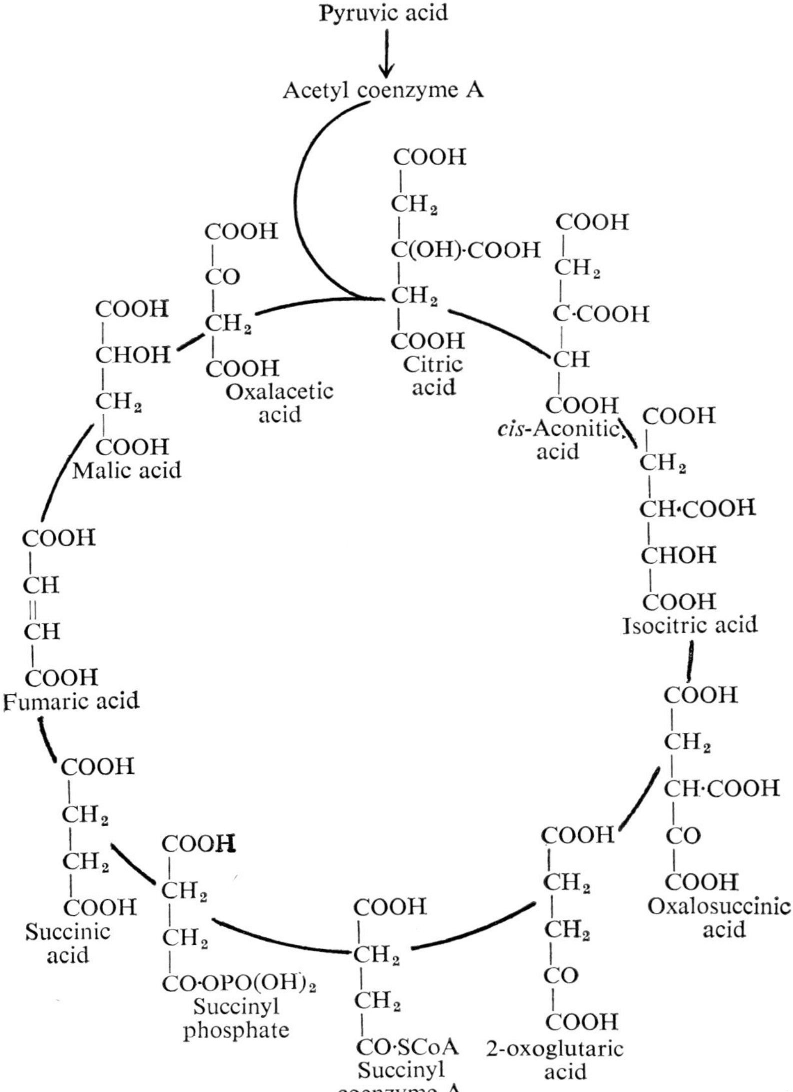

FIG. 24. Krebs' tricarboxylic acid or citric acid cycle

The citric acid so formed reacts as isocitric acid to form the next component in the cycle. An enzyme, aconitase, establishes an equilibrium between citric acid itself, *cis*-aconitic acid by elimination of the elements of water, and isocitric acid; there is thus available a pool of isocitric acid. The enzyme isocitric dehydrogenase (which acts in conjunction with NADP) oxidises isocitric acid by elimination of two hydrogen atoms, to oxalosuccinic acid, which in turn is converted by loss of carbon dioxide to 2-oxoglutaric acid; the enzyme catalysing this reaction is oxalosuccinic decarboxylase.

The next stage in the reaction is the conversion of 2-oxoglutaric acid to succinic acid. This is a complex reaction involving coenzyme A, NAD and guanosine triphosphate; 2-oxoglutaric acid loses carbon dioxide, and it is converted to succinyl-coenzyme A and thereafter to succinyl phosphate with regeneration of coenzyme A. Succinyl phosphate is hydrolysed to succinic acid with formation of GTP. Succinic acid is then oxidised to fumaric acid, a reaction catalysed by succinic dehydrogenase; a flavoprotein enzyme, fumarase, converts fumaric to malic acid which is in turn oxidised to oxaloacetate by malic dehydrogenase, an enzyme acting in conjunction with NAD. The oxaloacetate thus regenerated can re-enter the cycle, combining with more acetyl coenzyme A to re-form citric acid. The net reaction in this complex cycle is:

$$CH_3CO{\cdot}COOH + 2{\cdot}5O_2 = 3CO_2 + 2H_2O$$

This cycle yields energy to the cell.

In conclusion it should be pointed out that the Embden–Meyerhof tricarboxylic acid cycle is not the only route available to micro-organisms for the aerobic oxidation of carbohydrates. Glucose may be oxidised by at least three other pathways, the Entner–Doudoroff pathway, the hexose-monophosphate pathway or pentose pathway and one in which no phosphorylated intermediates are formed, called the direct oxidation pathway.

In some micro-organisms two paths may be used simultaneously. The nature and proportion of each path may be evaluated most conveniently by determining the distribution of a radioactive carbon isotope in the reaction products, starting with glucose labelled with radioactive carbon.

THE METABOLISM OF NITROGEN-CONTAINING COMPOUNDS

Although much emphasis is placed upon the study of carbohydrate breakdown by micro-organisms it should not be overlooked that the laboratory culture media most often used contains the breakdown products of protein only. Peptone water and the nutrient broth described on page 37 contain peptone and meat extract and it is a criterion of quality for these two products that they do not contain fermentable carbohydrate. Clearly, therefore, bacteria must be able to catabolise the nitrogenous constituents of these ingredients either aerobically or anaerobically and derive energy for growth thereby. A study of the metabolism of nitrogen-containing compounds is thus of great importance in understanding the overall picture of bacterial metabolism.

THE METABOLISM OF PROTEINS

Protein Breakdown

The process of putrefaction is due to the breakdown of protein, the principle nitrogeneous components of plants and animals, with the production of, amongst other products, evil-smelling amines. Many bacteria possess the ability to decompose proteins and the products derived from protein breakdown.

Proteolytic activity is of significance from a practical point of view in bacteriology in that it is used as a diagnostic tool. Many pathogenic bacteria produce toxins which are proteolytic enzymes and cause tissue damage by the exertion of this activity. Proteolytic activity can cause spoilage to foodstuffs composed of protein materials. Proteolytic activity is of importance in the nitrogen cycle in nature.

Proteins consist of an array of α-amino acids linked through their amino and carboxylic acid groups by elimination of water to form the peptide link.

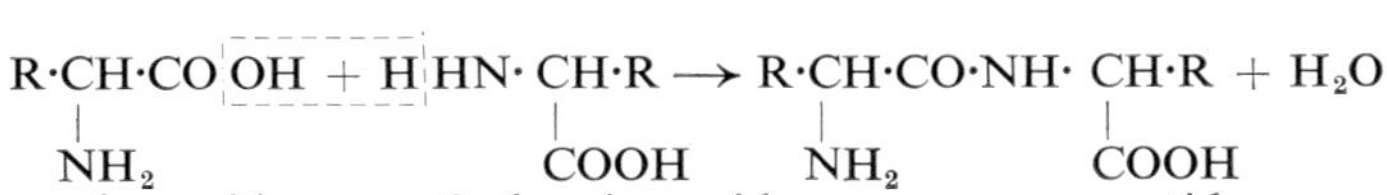

1st amino acid 2nd amino acid peptide

Protein decomposition thus must involve in the first instance the hydrolysis of this peptide link. A more fundamental understanding of the nature of proteolytic enzymes was made possible when a technique for synthesising peptides was developed.

Originally proteolytic enzymes were classified according to whether they decomposed proteins or protein subunits such as peptones and peptides. This idea has now been reformulated and proteolytic enzymes are classified as exopeptidases if they hydrolyse peptide bonds at or near the ends of polypeptides or proteins and endopeptidases if they are able to hydrolyse peptide bonds near the centre of polypeptides or proteins. In addition the activity of proteolytic enzymes was found to be affected by the nature of the amino acid adjacent to the peptide bond and by whether there was a free —COOH or —NH_2 group near the bond. Thus, amongst the exopeptidases are recognised the carboxypeptidases which are only active against peptide links provided that there is a free carboxyl group in the vicinity and the aminopeptidases which function only if there is a free amino group near the peptide bond concerned. A third type of exopeptidase, called a dipeptidase, is active against dipeptides.

Proteolytic activity in routine bacteriology is tested for by measuring the ability of organisms to liquefy gelatin or to digest clotted milk. Ability to liquefy gelatin is tested by inoculating a nutrient medium solidified with gelatin and, after incubation, chilling the medium in ice water and determining by inversion of the tube whether the gelatin has been liquefied. Hydrolysis of the peptide links of the macromolecular protein causes a progressive reduction in viscosity.

It should be remembered, however, that gelatin is an artificial product and may have undergone chemical change during its manufacture. In fundamental studies of protein breakdown by micro-organisms, a wider variety of substrates should be used, including natural or native protein (collagen is the natural protein from which gelatin is derived). Furthermore, evidence of the attack on protein should be sought by a determination of the free amino or carboxyl groups produced as the protein is hydrolysed.

Many proteolytic enzymes are extracellular and can be recovered from the culture medium in which the cells are grown. The production of proteolytic enzymes under artificial conditions by bacteria is affected by the composition of the medium in which the cells are grown, and the growth temperature. The main factors appear to be the correct balance of inorganic salts, of which a source of calcium ions seems to be the most important, and in some instances the presence of a utilisable source of carbohydrates.

Protein Synthesis

Protein synthesis occurs by the enzyme-catalysed elimination of water between successive pairs of amino acids, that is by a reversal of the protein degradation processes described under protein breakdown. The generation of peptide bonds, however, requires energy and it has been found that adenosine-triphosphate (ATP) is implicated in peptide bond and protein synthesis.

ATP reacts with amino acids to give a compound of adenosine-monophosphate (AMP) and the amino acid and a molecule of pyrophosphate is formed. This reaction is catalysed by a specific enzyme for each amino acid involved. The amino acid/AMP compound now combines with a small molecular weight RNA (the so-called soluble or transfer RNA) and this new complex, still containing the energy originally in the ATP, is transferred to the ribosome, which is the actual site of peptide bond and hence protein synthesis. At the ribosome the complex now reacts with a second amino acid forming a peptide and the AMP and RNA are available for further use by the cell. The amino acid sequence, which in turn controls the nature of the protein formed, is determined by a RNA molecule known as messenger or template RNA.

Many antibiotics act by interfering with protein synthesis.

Peptide Synthesis

A reaction known as transamidation or transpeptidation has been demonstrated in some bacterial species. The reaction may be illustrated by an example: glutamylcysteinylglycine + phenylalanine → glutamylphenylalanine + cysteinylglycine. Glutamylcysteinylglycine, a tripeptide, is known as glutathione and is thought to participate in many general reactions of this type.

THE METABOLISM OF AMINO ACIDS

It has been seen that the result of the action of proteolytic enzymes on proteins is the production of amino acids. Amino acids undergo a diversity of catabolic processes, the more important of which will be considered below. In addition, amino acids produced by the breakdown of proteins and peptides may be used by the cell for synthetic or anabolic processes. Catabolic reactions of amino acids may involve either the NH_2— group or the —COOH group.

Reactions of Amino Acids Involving the Amino Group

Oxidative deamination. This type reaction is catalysed by a group of enzymes called the amino acid oxidases. The result of the reaction is the conversion of an amino acid to an oxo acid and it has been proved in some cases that the reaction proceeds via an intermediary amino acid.

$$\begin{matrix} R \\ | \\ CH{\cdot}NH_2 \\ | \\ COOH \end{matrix} \xrightarrow[\text{oxidase}]{\text{Amino acid}} \begin{matrix} R \\ | \\ C{:}NH \\ | \\ COOH \end{matrix} \xrightarrow[+H_2O]{\text{Hydrolysis}} \begin{matrix} R \\ | \\ C{:}O \\ | \\ COOH \end{matrix} + NH_3$$

Thus the amino acid, glutamic acid, yields the oxo acid 2-oxoglutaric acid and alanine yields pyruvic acid.

The oxo acid produced is, under normal circumstances, metabolised as soon as it is formed, but by using special techniques may be isolated. Flavoproteins are the most widely encountered cofactors in bacterial amino acid oxidases. Oxidases exhibiting stereospecificity, that is active against only the D or L form of an amino acid, have been isolated.

The formation of indole. Tests for the formation of indole have long been used in diagnostic bacteriology and various techniques have been devised to demonstrate its presence. Indole arises from the breakdown of the amino acid tryptophan and its formation depends on the presence of this compound in the culture medium and the presence in the bacteria of the necessary enzymes. The reaction proceeds according to the following scheme.

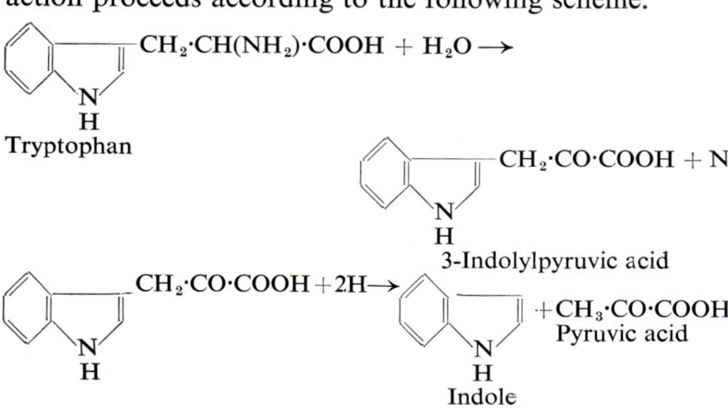

The overall reaction is catalysed by an enzyme complex called tryptophanase which has been found in various species of the bacterial genera *Escherichia*, *Proteus* and *Vibrio*. Tryptophanase has been the subject of much study, and it has been found that a high level of the enzyme can be induced if cells capable of developing tryptophanase activity are grown in the presence of tryptophan and under aerobic conditions. Pyridoxal-5′-phosphate has been shown to be necessary for tryptophanase activity.

Indole is detected in bacterial cultures by adding *p*-dimethylaminobenzaldehyde solution, when a characteristic pink colour is produced.

Not all bacteria produce indole from tryptophan; members of the genera *Bacillus* and *Pseudomonas* may break down this compound via a complex series of steps to 3-oxoadipic acid, the nitrogen being lost as ammonia.

Transamination. This reaction involves the exchange of an amino group between an amino acid and a keto acid and may best be illustrated by examples.

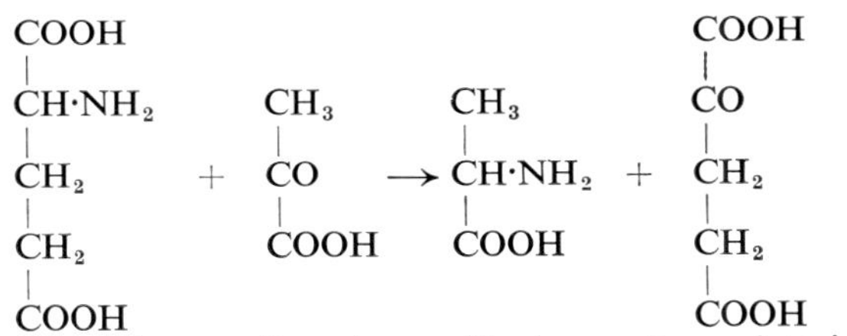

Glutamic acid | Pyruvic acid | Alanine | 2-oxoglutaric acid

Similarly:

Aspartic acid + pyruvic acid ⟶ oxaloacetic acid + alanine
Glutamic acid + oxaloacetic acid ⟶ 2-oxoglutaric acid + aspartic acid

In an extensive study with *Escherichia coli*, *Pseudomonas aeruginosa* and *Bacillus subtilis*, alanine, aspartic acid, leucine, methionine, norleucine, phenylalanine, tryptophan and tyrosine were shown to be able to transfer their amino group to 2-oxoglutaric acid with formation of glutamic acid. It should be appreciated that this reaction forms a route for the biosynthesis of amino acids as well as for their decomposition. All transaminases found so far in the bacteria are specific for the L-isomer of the amino acids involved and pyridoxal-5′-phosphate has been found to be the coenzyme for this reaction.

The Stickland reaction. Stickland, in 1934, showed that amino acids could take part in anaerobic oxidation-reduction reactions, certain acids acting as hydrogen donors and others functioning as hydrogen acceptors. The reaction was first studied in the anaerobic organism *Clostridium sporogenes*, and for this organism amino acids could be divided into two categories (Table 6).

TABLE 6

Hydrogen donor	Hydrogen acceptor
Histidine	Tryptophan
Alanine	Proline
Serine	Ornithine
Valine	Hydroxyproline
Tyrosine	Arginine
Leucine	Glycine

The reaction between alanine and proline was found to be as follows:

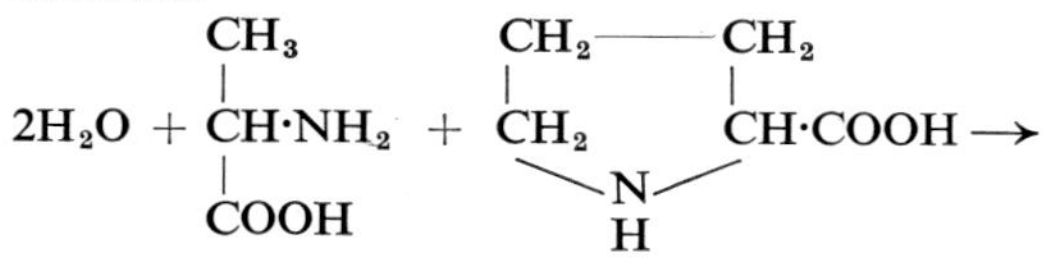

$$2CH_2NH_2{\cdot}(CH_2)_3{\cdot}COOH + NH_3 + CO_2 + CH_3COOH$$

5-amino valeric acid

The Stickland reaction yields energy to the cell and is an example of an anaerobic energy-yielding reaction.

Formation of hydrogen sulphide from sulphur-containing amino acids. Hydrogen sulphide production in culture media occurs in the case of heterotrophic bacteria as a result of the decomposition of sulphur-containing amino acids which themselves may be produced by the prior decomposition of proteins or peptones. Hydrogen sulphide production is used in diagnostic bacteriology and a culture medium containing an iron salt is often used to detect it, hydrogen sulphide production being indicated by the formation of black iron sulphides. Its formation may also be demonstrated by adding a thick washed suspension of, for example, *Proteus vulgaris* to a solution of the sulphur-containing amino acid cysteine in a small tube and trapping a piece of lead acetate paper in the cotton-wool plug of the tube. If hydrogen sulphide is evolved the lead acetate paper will be blackened within 3–4 hours after incubation.

In one mechanism for hydrogen sulphide production an enzyme called cysteine desulphydrase decomposes cysteine to hydrogen sulphide and α-aminoacrylic acid. The latter compound then hydrolyses spontaneously to form pyruvic acid and ammonia. Pyridoxal-5′-phosphate is the coenzyme for the desulphydrase.

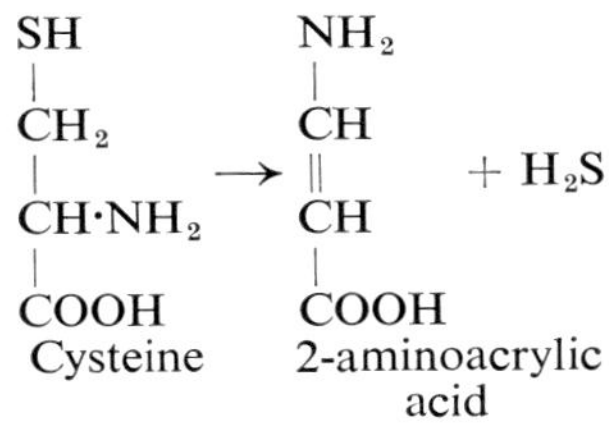

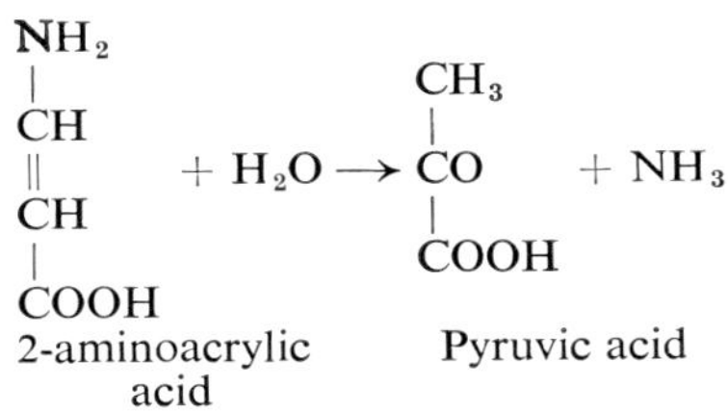

Reactions of Amino Acids Involving the Carboxylic Acid Group

The amino acid decarboxylases. This group of enzymes catalyses the following type reaction:

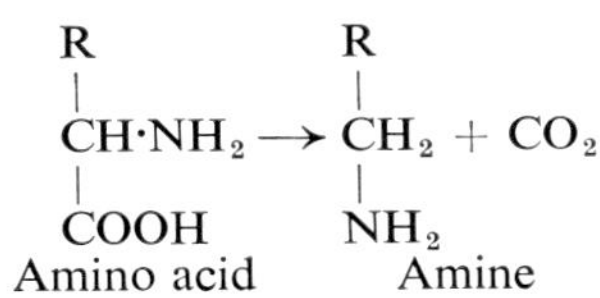

As can be seen, this reaction results in the production of an amine from a carboxylic acid; thus, on decarboxylation, histidine yields histamine, tyrosine yields tyramine and glutamic acid yields α-aminobutyric acid.

Gale made an extensive study of the bacterial amino acid decarboxylases. It was found that the enzyme preparations could be made in a stable form by drying the appropriate bacterial cells with acetone, that many decarboxylases were highly specific for a given amino acid, that the L-isomer only was attacked, and that a quantitative relationship existed between the amount of amino acid and the volume of carbon dioxide evolved. This last finding has made available a valuable analytical tool for determining individual amino acids in mixtures by measuring the volume of carbon dioxide evolved.

A most interesting relationship was also found to hold between the activity and the optimum pH of the amino acid decarboxylases. This family of enzymes had their pH optima in the acid range, pH 4–5. In contrast, the deaminases were most active in the alkaline range, pH 7·5–8. It has been suggested that this phenomenon is a natural neutralisation mechanism. Thus, if in a culture of bacteria the pH becomes acid, then the amino acid decarboxylases, if present, are at their most active and will decarboxylate any amino acids present forming amines, and carbon dioxide will be evolved. These amines will tend to neutralise the medium. Conversely, if the reaction becomes alkaline, amino acids will be deaminated with production of a carboxylic acid, although the highly soluble ammonia evolved may dissolve in the medium and deamination may not be such an effective neutralising reaction as decarboxylation. Pyridoxal-5′-phosphate has been implicated as a coenzyme in this group of enzymes.

THE METABOLISM OF NUCLEIC ACIDS

Nucleic acids are polymers, the repeating units of which are called nucleotides. Nucleotides themselves consist of a base, a sugar and orthophosphoric acid. The sugar is either ribose or deoxyribose, giving rise to ribonucleic acid (RNA) or deoxyribonucleic acid (DNA).

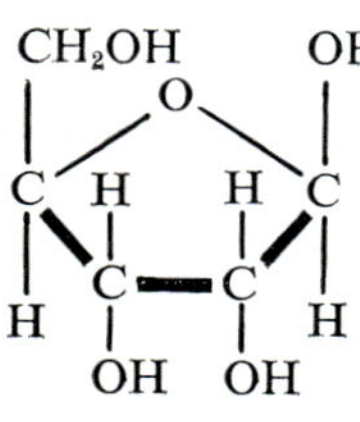

Ribose

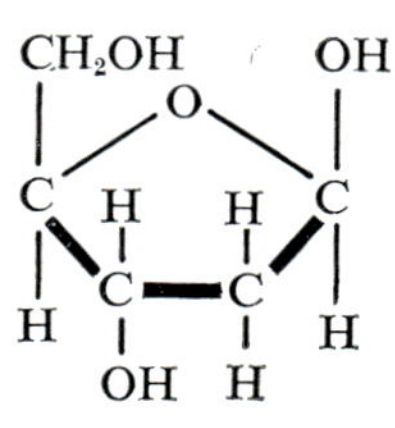

Deoxyribose

The bases found in the natural nucleic acids are derivatives of pyrimidine and purine and are constituted as follows: adenine (6-aminopurine), guanine (2-amino-6-hydroxypurine), uracil (2,6-dihydroxypyrimidine) and cytosine (6-amino-2-hydroxypyrimidine) in RNA and adenine, guanine, cytosine and thymine (2,4-dihydroxy-5-methylpyrimidine) in DNA, together

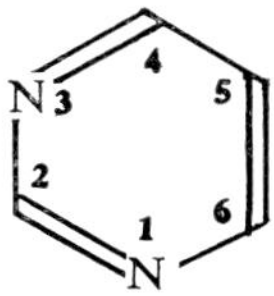

Pyrimidine

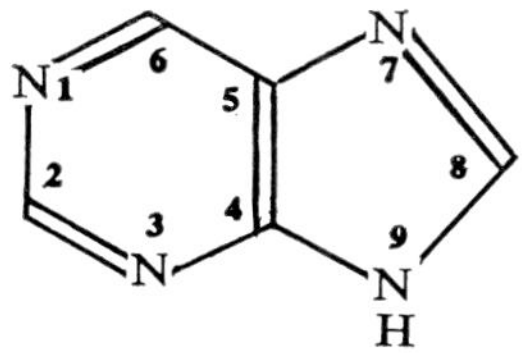

Purine

with nicotinamide (pyridine-3-carboxylic acid amide) in many co-enzymes, and uracil in transglycosidases. A nucleotide may be represented thus:

Orthophosphoric acid + Sugar + Base

Sugar + Base = Nucleoside

Orthophosphoric acid + Nucleoside = Nucleotide

The nucleotide chains of DNA are arranged in the form of a double spiral cross-linked by hydrogen bonding between adjacent bases, as has recently been demonstrated by the Nobel Prize winners Crick and Watson (Fig. 20). This spiral template is responsible for the handing on of genetic information in the process of cell division. DNA is found associated with the nuclear apparatus in the bacterial cytoplasm but, as stated on page 14, is not enclosed within a membrane to give the typical nucleus as in other living cells.

It will be recalled that two forms of RNA have been recognised, a small molecular weight or soluble RNA, and a normal, or by contrast, macromolecular weight RNA which is found in the ribosomes. Both have been shown to be closely linked with the biosynthesis of proteins but are not concerned with heredity. More recently soluble RNA because of its function in the cell has been called transfer RNA.

The following coenzymes have a nucleotide structure: the adenosine phosphates, NAD and NADP, the flavin nucleotide, flavin-adenine dinucleotide (FAD), coenzyme A and the uracil coenzymes involved in transglycosidisation. Uracil-containing nucleotides are involved in the synthesis of the rigid component of bacterial cell walls and cell wall precursors, and accumulate in the medium when bacteria are treated with penicillin and other inhibitors of cell wall synthesis.

The enzymes which degrade nucleic acids are summarised in Table 7. The pentose sugars may be further degraded or utilised for other synthetic processes. The bases produced may themselves

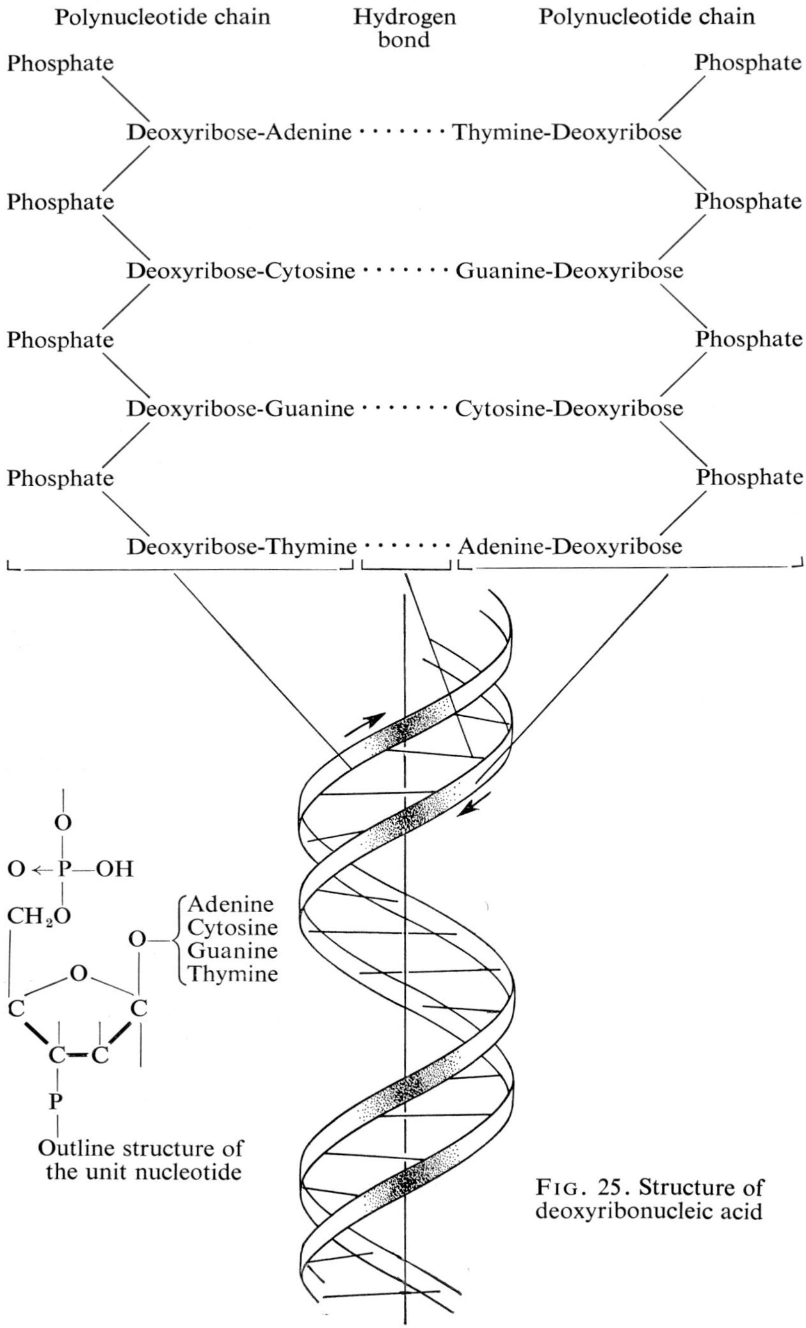

FIG. 25. Structure of deoxyribonucleic acid

TABLE 7

ENZYMES WHICH DEGRADE NUCLEIC ACIDS

Substrate	Enzyme	Product
Ribonucleic acid	Ribonuclease	Nucleotide
Deoxyribonucleic acid	Deoxyribonuclease	Nucleotide
Nucleotide	Nucleotidase	Phosphoric acid + Nucleoside
Nucleoside	Nucleosidases	Ribose or deoxyribose base

undergo further metabolism. In some cases bacteria have been isolated which can utilise the purines and pyrimidines as sole sources of carbon. In general it has been found that the purines are more readily decomposed than the pyrimidines. For example, purines are degraded by species of the bacterial genera *Clostridium* and *Micrococcus*, and carbon dioxide, hydrogen, acetate, lactate, glycine and ammonia have been identified as the end-products. *Micrococcus aerogenes* can decompose the pyrimidines uracil, cytosine and thymine to carbon dioxide, hydrogen, ammonia, lactate and acetate; while an organism called *Clostridium uracilicum* decomposes uracil to alanine, carbon dioxide and ammonia.

One organism has been described, a so far unidentified clostripium, which decomposes nicotinamide to give ammonia, acetate, propionate and carbon dioxide.

THE REDUCTION OF NITRATE

Many bacteria can reduce nitrates to nitrites, nitrous oxide, nitrogen gas, hydroxylamine or ammonia. These several abilities may be tested by growing the organisms in a nutrient broth containing 0·1 per cent potassium nitrate and a fermentation tube. After growth the possible reduction products of nitrate are tested for as follows.

Nitrite. Add sulphanilic acid in acetic acid followed by α-naphthylamine acetate. If nitrite is present the sulphanilic acid is diazotised and will now couple with the α-naphthylamine to give a magenta colour. This test can be adapted to give a quantitative estimation of nitrite production by micro-organisms.

Nitrogen gas. This may be detected visually by observing the collection of gas in the fermentation tube.

Nitrous oxide. This will also collect in the fermentation tube, if evolved in excess of its solubility.

Hydroxylamine. This may be reoxidised to nitrite with iodine, and the nitrite detected and estimated as above.

Ammonia. Test with Nessler's reagent.

The Mechanism of Nitrate Reduction

Nitrite, nitrous oxide, and nitrogen formation. The essential stage in the formation of nitrite or nitrogen may be written as follows:

$$2NO_3' \rightarrow 2NO_2' + O_2$$
$$2NO_3' \rightarrow N_2 + 3O_2$$

It is clear that there must be present in the reaction mixture a system capable of reacting with and reducing oxygen, and as a direct consequence of this, nitrate may act as an oxidising agent or hydrogen acceptor in cellular metabolism.

The presence of molecular oxygen tends to suppress nitrite formation. Thus in *Esch. coli* the presence of only 1 per cent oxygen will reduce nitrite formation to 40 per cent of the value in its absence; 20 per cent oxygen leads to 6 per cent conversion of nitrate to nitrite only. Thus nitrate reduction tested for as described above and carried out in open test tubes depends on the degree of anaerobiosis achieved by the inoculum and the extent of diffusion and solution of atmospheric oxygen into the medium during the course of the test.

The pathways, by which the various reduction products of nitrite (nitrous oxide and nitrogen) are formed, are complex, and intermediates such as nitramide and hyponitrous acid have been postulated.

Ammonia production. Many bacteria can reduce nitrate to ammonia; hydroxylamine is thought to be an intermediate in this reduction, which is less sensitive to oxygen than are the reductions mentioned above.

THE METABOLISM OF AROMATIC RINGS

Some bacteria, of which members of the genus *Pseudomonas* predominate, are able to decompose such compounds as phenol, catechol and benzoic acid, producing eventually 3-oxoadipic acid, which is converted in turn to succinic and acetic acid and is thus oxidisable via the Krebs' cycle.

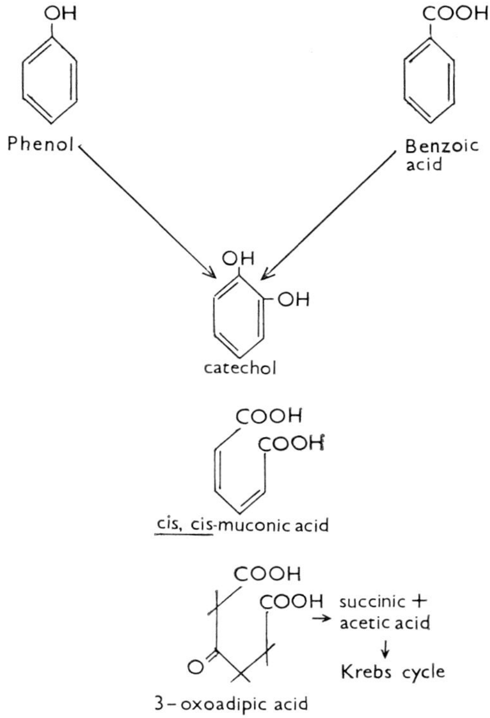

THE METABOLISM OF FATS

The fats are long chain fatty acid esters of glycerol, and are usually first hydrolysed by lipases to yield the fatty acid and glycerol, and this reaction contributes to the rancid flavour found in fats and oils which have undergone spoilage by micro-organisms.

The glycerol may be further metabolised by the reaction sequences set out in the Embden–Meyerhof and Krebs' cycles. The fatty acid is converted to a derivative of coenzyme A, which then undergoes oxidation in a series of reactions whereby two carbon atoms are successively lost, giving finally acetylcoenzyme A, which can then be oxidised to water and carbon dioxide in the Krebs' cycle.

Some micro-organisms are able to synthesise fats and it is thought that the above reactions working in reverse or reactions involving malonyl coenzyme A are the routes by which this synthesis occurs. In man, fat synthesis does not proceed via a reversal of the catabolic process.

The steps in this reaction are as follows. The fatty acid is first converted to the corresponding acyl-coenzyme A this is then converted to a dehydro or unsaturated acyl coenzyme A, the two hydrogens lost with the generation of a double bond occurring immediately before the acyl-CoA grouping i.e. $CH_3(CH_2)_n{\cdot}COOH$ becomes $CH_3(CH_2)_{n-1}{\cdot}CH{=}CH{\cdot}COOH$. Next, the unsaturated fatty acid becomes hydroxylated to yield

$$CH_3(CH_2)_{n-2}{\cdot}CHOH{\cdot}CH_2{\cdot}COOH$$

by addition of water across the double bond of the unsaturated fatty acid; this hydroxy group is now oxidised to a keto or oxo group to give $CH_3(CH_2)_{n-2}{\cdot}CO{\cdot}CH_2{\cdot}COOH$. This oxo acid now decomposes to give the fatty acid with two carbon atoms less than the starting acid and acetic acid as acetyl-coenzyme A the latter accounting for the two carbon loss.

Again it must be emphasised that all these intermediates and the products of the final reaction react as derivatives of coenzyme A. This is a continuous process and the fatty acid containing now two carbon atoms less than the starting acid re-enters the reaction sequence.

The process has been named the fatty acid spiral and may be represented in 'shorthand' thus:

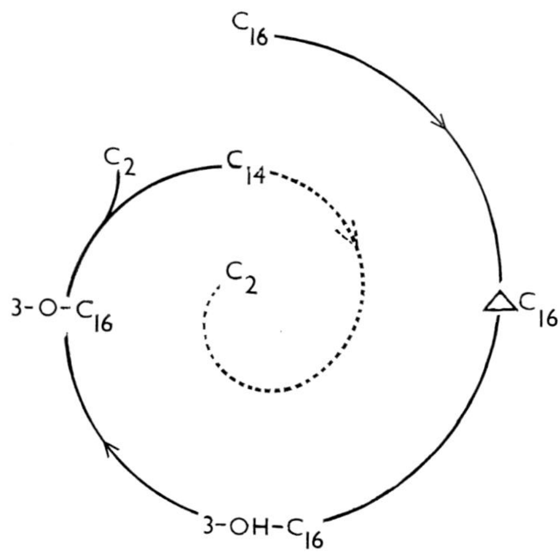

C_{16}, C_{14}: fatty-acyl-coenzyme A; C_2: acetyl-coenzyme A; ΔC16: unsaturated fatty-acyl-coenzyme A; 3-OH-C_{16}: 3-hydroxyacyl-coenzyme A; 3-O-C_{16}: 3-oxoacyl-coenzyme A.

CHEMICALS USED IN METABOLIC STUDIES

Sodium fluoride. In the presence of phosphate, fluoride forms an insoluble magnesium fluorophosphate and will thus inhibit enzymes which require magnesium for their activity. An example of such an enzyme is enolase.

Iodoacetic acid. This substance is an oxidising agent and acts on enzymes which require thiol (—SH) groups for their activity. Examples of such enzymes include the glyceraldehyde-3-phosphate dehydrogenase, and enzyme reactions involving coenzyme A.

Sodium hydroxymercuribenzoate. This reagent is also highly specific for cellular thiol groups and may be used to determine them quantitatively.

Sodium arsenite. This compound is also a reagent for thiol groups.

Sodium arsenate. Competes for phosphate in biological phosphorylations. It is thus a phosphorylation inhibitor.

Sodium azide. This compound has a complex action on cellular metabolism. It appears in certain cases to inhibit the formation of high-energy phosphate bonds and also to poison iron-containing enzyme systems.

Cyanides. The soluble cyanides combine with iron-containing enzymes and thus inhibit cellular respiration involving the cytochrome system.

Carbon monoxide. This is also a specific inhibitor of iron-containing enzymes. It is a competitive inhibitor, while cyanide is a non-competitive inhibitor.

Dinitrophenol. This compound is able to inhibit the formation of high-energy phosphate bonds. It is sometimes spoken of as an uncoupling reagent, in that it can, when added to a cellular system carrying out oxidative phosphorylation, uncouple or disconnect the two interlinked processes, preventing the synthesis of phosphate esters containing the so-called high bond energy (free energy of hydrolysis) but allowing the oxidative process such as oxidation of reduced NAD, NADP, FAD, or cytochromes or substrates such as glucose or pyruvic acid to proceed unhindered.

References

The Bacteria. Vol. 2: *Metabolism*. Ed. Gunsalus, I. C., and Stanier, R. V. Academic Press, 1961

Metabolic Maps. Vol. 2. Umbreit, W. W. Burgess Publishing Co., Minneapolis, 1960

Enzymes. 2nd Ed. Dixon, M., and Webb, E. C. Longmans, Green and Co., London, 1964

CHAPTER 4

GENERAL PROPERTIES OF MICRO-ORGANISMS OF SIGNIFICANCE IN PHARMACY

THE CLASSIFICATION AND CHARACTERISTICS OF IMPORTANT BACTERIA

Bacteria have been traditionally classified as a sub-division of the fungi but the same microscopical, biochemical and biophysical examinations which gave rise to the detailed knowledge outlined in Chapter 1 have necessitated a reappraisal of the place of the bacteria in the family tree of living organisms, and bacteria have been shown to possess, in company with the blue-green algae, a series of very different properties not shared by all other living organisms. Among these are included a nucleus not enclosed in a nuclear membrane, the possession of peptidoglycan as a cell wall component, ribosomes differing in their weight, as demonstrated by differential centrifugation, of their sub-units, the absence of mitochondria and certain other properties. As a result, the bacteria and blue-green algae have been designated procaryotes or procaryotic organisms while all other organisms from green algae to man are called eucaryotic organisms or eucaryotes. The term caryote refers to the nucleus and the two divisive terms mean possessing a 'primitive' or 'first evolving' nucleus of a 'true' nucleus. As has been seen, the possession or otherwise of a nuclear membrane is accompanied by a number of other fundamental differences. These are summarised and extended in Table 8.

The further classification of bacteria, blue-green algae and the remaining living organisms is not affected by this division, and remains as now extant and in more conventional terms the organisms separated by the data given above appear thus:

Schizophyta	Thallophyta
1. Bacteria	1. Algae
2. Blue-green algae	2. Fungi
	3. Lichens

It is not easy to fit the ricketsiae, viruses and bacteriophages

TABLE 8

	Procaryotes	Eucaryctes
Peptidoglycan in cell wall	Present	Absent
Cellulose in cell wall	Absent	Present
Teichoic acids	Present	Absent
Nucleus	Non-membrane bound	Membrane bound
Mitotic division	Absent	Present
Mitochondria	Absent	Present
Mesosomes	Present	Absent
Chloroplasts	Absent	Present
Photosynthetic vesicles in cytoplasm	Present	Absent
80*S* ribosomes	Absent	Present
70*S* ribosomes	Present	Absent
Ability to fix nitrogen	Present	Absent
Non-glycolytic energy pathways	Present	Absent
Use of inorganic compounds as energy source	Present	Absent

into the above scheme, and they have accordingly been omitted from it.

The classification of bacteria presents today an unstable and changing picture, despite the time that has elapsed since the first attempts at a classification of the known bacteria were made by Ferdinand Cohn in 1872.

In this chapter the properties and cultivation of the more important organisms encountered in medicine and pharmacy will be considered.

The standard work which treats the subject in great detail is Bergey's *Manual of Determinative Bacteriology*. In the seventh edition of this book the bacteria are divided into 10 orders. Of these, only four contain organisms of interest in pharmacy and medicine. All references to this manual refer to the seventh edition. Bacterial orders are subdivided further into sub-orders, families, genera and species. Species may be further subdivided into strains. Strains within a species may only differ from one another by one character.

Bergey order number	Order name
I	Pseudomonadales
IV	Eubacteriales
V	Actinomycetales
IX	Spirochaetales

Identification of Bacteria

Methods for the identification of the common medical bacteria have recently been described by Drs Cowan and Steel, who have grouped in convenient tables many of the biochemical reactions of the bacterial genera and species described in the following pages. This publication (see references) is recommended to all students as an interesting assembly of the essential laboratory tests for identifying common bacteria and for a useful collation of their main diagnostic properties.

ORDER: PSEUDOMONADALES

This order is divided into two sub-orders. The first contains photosynthetic bacteria, the second non-photosynthetic organisms.

Within the second sub-order there are seven families, two of which contain a genus of interest in pharmacy and medicine.

	Pseudomonadales		*Order*
Rhodobacteriales	Pseudomonadineae		*Sub-orders*
	Pseudomonadaceae	Spirillaceae	*Families*
	Pseudomonas	Vibrio	*Genera*

Genus *Pseudomonas*

Members of this genus are almost invariably straight rods 1·5 to 3 μm in length. They are non-sporing and Gram-negative and possess either a single or many terminal flagella. Another characteristic of the genus is the production of extracellular yellow, green or blue pigments, some of which are strongly fluorescent, although under certain conditions of culture this feature may be lost.

Many of the members of this group are plant pathogens. Others are pathogenic to man and animals.

Pseudomonas aeruginosa, an important species which can infect wounds, is extremely difficult to eradicate as it is resistant to most antiseptic and antibiotic treatment. The organism is often referred to as *Pseudomonas pyocyanea* (the second, specific, name meaning blue pus), as it often causes a characteristic blue colour in pus when it has infected a wound, due to the production of pigments. It also infects the alimentary tract causing diarrhoea in infants. *Ps. aeruginosa* is an aerobe which grows best at 30°–37° C and may be cultivated on conventional nutrient broth: it is nutritionally non-exacting. It produces acid but not gas from glucose, and gives negative results in the Voges–Proskauer test. It does not produce indole from tryptophan, reduce nitrate to nitrite or produce hydrogen sulphide from sulphur-containing amino acids. Many species can decompose aromatic compounds and utilise them as the sole carbon source.

Pseudomonas fluorescens differs from *Ps. aeruginosa* in that it does not produce the blue pigment (pyocyanin) although it produces fluorescent pigments; it grows best at 25° C and is less pathogenic to animals than *Ps. aeruginosa*. It is, however, pathogenic to many plants.

Genus *Vibrio*

The cells of this genus occur often as curved or spirally twisted rods, although this character is not always present. Organisms are Gram-negative and flagellated.

Vibrio comma, discovered by Koshim in 1883, is the causal organism of Asiatic cholera. In man, the symptoms of cholera may vary from an attack of diarrhoea to general dehydration leading to death within 12 hours after infection. The last epidemic of cholera in Britain was in 1866, but the disease is endemic in India and Burma. Cholera is often water-borne, and may be contracted by drinking infected water but also by eating infected food or by direct contact with the stools or vomitus of infected persons.

V. comma grows on ordinary culture medium. It is particularly sensitive to acid, and by contrast grows well in an alkaline environment.

A medium consisting of alkali-treated defibrinated ox blood, solidified with agar, and having a pH of 9·0–9·6 (Dieudonné's medium) is used for its cultivation. The alkaline reaction helps to prevent the growth of other organisms when *V. comma* is being sought in faeces, water or foodstuffs.

Cholera vaccine is used as a prophylactic agent for persons who may be exposed to the disease.

ORDER: EUBACTERIALES

This order is divided into 13 families of which seven contain genera of interest in pharmacy (Table 9).

FAMILY ENTEROBACTERIACEAE

Members of this family are typically Gram-negative non-spore-forming rods. Some are flagellated and motile; many produce extracellular polysaccharide slime. The Enterobacteriaceae are biochemically active and ferment many sugars, producing acid or acid and gaseous end-products. They vary widely in their nutritional requirements. Many inhabit the intestinal tract of man and animals. Some are human and animal pathogens.

Genus *Escherichia*

This genus contains four species of which the most important member is *Escherichia coli*. This organism is typically rod-shaped

TABLE 9

ORDER: EUBACTERIALES

Family	Genus
1. Enterobacteriaceae	(*a*) *Escherichia* (*b*) *Aerobacter* (*c*) *Klebsiella* (*d*) *Serratia* (*e*) *Proteus* (*f*) *Salmonella* (*g*) *Shigella*
2. Brucellaceae	(*a*) *Pasteurella* (*b*) *Bordetella* (*c*) *Brucella* (*d*) *Haemophilus*
3. Micrococcaceae	(*a*) *Micrococcus* (*b*) *Staphylococcus*
4. Neisseriaceae	(*a*) *Neisseria*
5. Lactobacillaceae	(*a*) *Diplococcus* (*b*) *Streptococcus* (*c*) *Leuconostoc* (*d*) *Lactobacillus*
6. Corynebacteriaceae	(*a*) *Corynebacterium*
7. Bacillaceae	(*a*) *Bacillus* (*b*) *Clostridium*

(1 μm $\times$ 4 μm), Gram-negative, non-spore-forming and may be motile or non-motile. Motile forms are peritrichously flagellated. It can grow aerobically or anaerobically and ferments a large range of carbohydrates; it is nutritionally non-exacting and grows best at 30°–37° C; it reduces nitrates to nitrites, and produces indole from tryptophan. *Esch. coli* is found in the human intestine and in the intestine of many animals; its specific name is derived from the Latin 'coli'—from the colon. Because of its biochemical activity, the ease by which it can be grown and its relative lack of pathogenicity it has been much used for biochemical studies. Because of its habitat in the intestine its presence is used as a marker or indicator of faecal or sewage contamination in drinking waters; in general, water containing from 1 to 5 organisms of intestinal origin per 100 ml is considered unfit for drinking. Because of the

need to distinguish *Esch. coli* from other organisms in the gut and faeces which might be of greater danger to public health (organisms that cause typhoid fever, bacillary dysentery, for example), a variety of special culture media have been evolved to select and distinguish members of this group of organisms.

Genus *Aerobacter*

The member of this genus of interest, *Aerobacter aerogenes*, is in many ways similar to *Esch. coli*. However, its optimum growth temperature is 30° C and it tends to produce extracellular slime. It can also fix atmospheric nitrogen. It is not confined to the gut but is found widely distributed in nature. Because of this, tests have been evolved, based on differences in biochemical capabilities, to distinguish *Aerobacter aerogenes* from *Esch. coli*.

If a Gram-negative, lactose-fermenting, rod-shaped organism is isolated from water, it is obviously of importance to know if it is *Esch. coli* and an indication of faecal contamination or whether it is *A. aerogenes* which is not generally considered as evidence of faecal contamination (see page 127).

Genus *Klebsiella*

The species *Klebsiella pneumoniae* is one of the causes of pneumonia in man and can cause mastitis in cattle. It consists of short, non-motile, Gram-negative rods often invested with a capsule of carbohydrate material.

Genus *Serratia*

Serratia marcescens consists of short (0·5–1·0 μm) motile, Gram-negative rods. A characteristic of this organism is its production of a red pigment, prodigiosin. *Serr. marcescens* is used to test filters used for removing bacteria from solutions (see *B.P.C.*, 1968, p. 1421). The small size of this organism makes the test a severe one.

Genus *Proteus*

Members of this genus are rod-shaped, Gram-negative, non-spore-forming organisms. They are peritrichously flagellated, do not form capsules, and grow aerobically and anaerobically.

When inoculated on to solid media some species, especially if the surface of the medium is moist, tend to swarm or spread over the surface. This phase is associated with a change in the morphology of the organism which may reach a length of 30 μm (the normal organism is 1–3 μm long). Swarming occurs in phases, typically, a fresh burst of swarming starts every 3–4 hours. When the surface of the plate is covered the phenomenon ceases and the organisms revert to their original size. All members of this genus decompose urea to ammonia and carbon dioxide. *Pr. mirabilis* and *Pr. morganii* can infect the urinary tract of man. Because of their avid ability to decompose urea they produce alkaline conditions in the bladder; eradication from the genito-urinary tract is often difficult.

One biochemical capability of this genus which distinguishes it from the other seven genera of the Enterobacteriaceae is the ability to deaminate phenylalanine, converting it to phenylpyruvic acid. Phenylpyruvic acid is detected by the green colour it gives with ferric salts.

Genus *Salmonella*

Bergey's manual lists 10 species of this organism and 343 types which have been distinguished by serological methods. The first member of this genus to be described was the causal organism of typhoid fever, called *Salmonella typhosa* in Bergey's manual and *Sal. typhi* in most English texts.

Sal. typhosa is a Gram-negative, non-sporing rod, usually motile by peritrichous flagella although a few varieties are non-motile. It does not produce gas in fermentations as it lacks the enzyme systems necessary to decompose formic acid and unlike *Esch. coli* cannot ferment lactose. It grows well on ordinary culture media at 37° C. It is the test organism used in the Rideal–Walker and Chick–Martin tests for disinfectant evaluation. Typhoid fever may be spread by contaminated drinking water, foodstuffs or direct contact with the faeces of infected persons. The organisms can survive in public water supplies and the importance of detecting evidence of sewerage or faecal contamination in drinking waters is thus apparent. Foodstuffs can be contaminated during handling by infected persons unless rigorous personal hygiene is observed.

In some instances the organism may occur in the intestinal tract of a human subject without the latter showing any symptoms of the disease. Such a person can, however, transmit the organism as outlined above and is called a typhoid carrier.

In addition to typhoid fever a milder but similar condition known as paratyphoid fever is recognised. Paratyphoid fever is caused by *Sal. paratyphi* (*Sal. paratyphi* A), *Sal. schottmülleri*

(*Sal. paratyphi* B) and *Sal. hirschfeldii* (*Sal. paratyphi* C); the names in brackets are those used to describe the organisms in the monographs on vaccines containing these organisms in the *British Pharmacopoeia*. These three organisms are very similar in general character to *Sal. typhosa* but may be distinguished from one another and from *Sal. typhosa* by biochemical tests.

Certain salmonellae are the cause of one type of food poisoning. Two species, *Sal. typhimurium* and *Sal. enteritidis*, and many other types which have been identified by serological methods have been implicated in salmonella food poisoning. When the organisms invade the gastro-intestinal tract, endotoxins, released when a proportion of the cells undergo lysis, set up an irritation giving rise to nausea, vomiting and diarrhoea. The onset of these symptoms is usually from 12 to 60 hours after the contaminated food was eaten.

Genus *Shigella*

Members of this group are non-motile, non-sporing, Gram-negative rods, which decompose carbohydrates to produce acid products only without the additional gaseous end-products produced by some members of the family.

The Shigellae are the cause of acute bacillary dysentery in man. Bergey's manual describes eight species of which four, *Sh. dysenteriae*, *Sh. flexneri*, *Sh. sonnei* and *Sh. boydii*, are usually encountered in cases of dysentery.

Laboratory Differentiation of the Enterobacteriaceae

A very great number of ingeniously formulated media for the rapid differentiation of members of the Enterobacteriaceae have been devised. Such media are used in hospital and public health laboratories for identifying the organisms found in samples believed to be contaminated by these organisms, and as an aid to diagnosis and treatment. As examples, two such media will be described.

MacConkey's medium. This medium was introduced in 1905 to isolate Enterobacteriaceae from water, urine, faeces, foods, etc. Essentially it consists of a nutrient medium with bile salts, lactose, and a suitable indicator. The bile salts function as a natural surface active agent which, while not inhibiting the growth of the Enterobacteriaceae, inhibits the growth of Gram-positive bacteria which are likely to be present in the material to be examined.

Esch. coli and *A. aerogenes* produce acid from lactose on this medium, which alters the colour of the indicator, and also adsorb some of the indicator which may be precipitated around the growing cells. The organisms causing typhoid and paratyphoid fever and bacillary dysentery do not ferment lactose and colonies of these organisms appear transparent.

Many modifications of MacConkey's medium exist; one employs a synthetic surface active agent in place of bile salts.

Bismuth sulphite agar. This medium was developed in the 1920's for the identification of *Salmonella typhosa*, in water, faeces, urine or foods. It consists of a buffered nutrient agar containing bismuth sulphite, ferrous sulphate and brilliant green.

Esch. coli (which is also likely to be present in material to be examined) is inhibited by the concentration (0·0025 per cent) of brilliant green used, while *Sal. typhosa* will grow luxuriantly. Bismuth sulphite also exerts some inhibitory effect on *Esch. coli.*

Sal. typhosa, in the presence of glucose, reduces bismuth sulphite to bismuth sulphide, a black compound; this organism can produce hydrogen sulphide from sulphur-containing amino acids in the medium which will react with ferrous sulphate to give a black deposit of ferrous sulphide (Table 10).

TABLE 10

APPEARANCE OF BACTERIAL COLONIES ON BISMUTH SULPHITE AGAR

Organism	Appearance of colonies on bismuth sulphite agar
Sal. typhosa	Black with blackened extracolonial zone
Sal. enteritidis	Black with blackened extracolonial zone
Sal. schotmülleri	Black with blackened extracolonial zone
Sal. paratyphi *Sal. typhimurium* *Sal. choleraesuis*	Green
Sh. flexneri *Sh. sonnei*	Brown
Other shigellae *Esch. coli*	No growth

Esch. coli and *A. aerogenes* may be distinguished by their differing ability to produce indole when grown in peptone water, by their action on glucose in peptone water buffered at pH 7·5 and containing methyl red, and by their ability to utilise citric acid as sole carbon source. As explained in Chapter 3, *Esch. coli* produces acid products from glucose in sufficient quantity, that, despite the buffer capacity of the medium, the pH is reduced to below 4·2 and the methyl red indicator is turned from yellow to red—such a result is termed methyl red positive. *A. aerogenes* produces acetylmethylcarbinol from glucose; this product is not acid, the pH is not reduced to below 4·2, and the methyl red indicator remains yellow or orange. This organism is described as methyl red negative. The acetylmethylcarbinol can be detected by suitable reagents (Voges–Proskauer (V.P.) reaction). *A. aerogenes* is V.P. positive.

	Indole	MR	VP	Citrate
Esch. coli	+	+	−	−
A. aerogenes	−	−	+	+

In summary, therefore, the behaviour of the two organisms is as given above.

The four tests are distinguished by a mnemonic IMViC, and many texts refer to the IMViC characteristics of these and other related organisms.

The differing ability to ferment sugars, glycosides and polyhydric alcohols is widely used to differentiate the Enterobacteriaceae and in diagnostic bacteriology generally. The test is usually carried out by adding the reagent aseptically to sterilised peptone water and a suitable indicator contained in a small (5 ml) bottle closed with a rubber-lined screw cap and containing a small inverted tube filled with the medium. Acid production is indicated by a change in colour of the indicator, and gas production by gas collecting in the inverted tube. A more informative test method is described by Hugh and Leifson (see references).

FAMILY BRUCELLACEAE

Members of this family are Gram-negative, motile or non-motile, non-sporing, short rods. Some forms are coccoid. They will grow aerobically or anaerobically and are nutritionally exacting.

Genus *Pasteurella*

The plague or Black Death as it was called during the epidemic of 1348 is caused by a member of this genus, *Past. pestis*, which also infects rats and is transmitted to man and spread in rat colonies by the rat flea. Two main forms of plague are recognised, one, infecting the lymphatic system and giving rise to swellings or bubos in the lymph glands, is called bubonic plague, a second rapidly fatal form affects the respiratory system and is called pneumonic plague.

Past. pestis is a Gram-negative, non-spore-forming cocco-bacillus which when stained is more intensely coloured at the ends of the cell (bipolar staining). The organism is biochemically inert and differs from most human and animal pathogens in that its optimum growth temperature is some ten degrees below body temperature. It will grow both aerobically and anaerobically on ordinary culture media. The organisms are capsulated, and this capsule is associated with the pathogenicity of the organism. A vaccine is described in the *British Pharmacopoeia* where it is specified that the final vaccine contains the greatest possible amount of capsular material.

Past. tularensis is the cause of tularaemia in man, and is almost invariably contracted by handling animals infected with the disease. The disease is endemic amongst many species of wild rodents in the American mid-west.

Genus *Bordetella*

This genus contains the causal organisms of whooping-cough. *Bordetella pertussis*, a small, oval, Gram-negative organism, which does not form spores and is non-motile, is aerobic, biochemically inert and exacting in its nutritional requirements.

A vaccine is described in the *British Pharmacopoeia* for the prevention of whooping cough. Vaccine therapy has been highly successful in eradicating this complaint or reducing the severity of attacks.

Genus *Brucella*

This genus contains the organism *Brucella melitensis*, the causal organism of Malta fever, an undulant fever seldom met in this country except amongst people who have contracted the disease abroad. *Brucella bronchisepticus* is used for the microbiological assay of the antibiotic polymyxin B.

Genus *Haemophilus*

The generic name of this group means blood-loving and arose because these organisms could not be grown on artificial media without the addition of whole blood. It has been found that two constituents from blood, a nicotinamide nucleotide and haematin, are responsible for its growth-promoting properties. Further research amongst members of the genus has shown that some members require both these factors, and some, one or the other; this fact is used to distinguish species.

One member, *Haemophilus influenzae*, was at one time regarded as the cause of influenza in man. It is often found in the respiratory tract during attacks of influenza, the cause of which is now established as being due to a virus. *H. influenzae* can infect the respiratory tract causing inflammation however.

H. influenzae consists of short, Gram-negative, non-motile rods which exhibit bipolar staining. It requires both a nicotinamide nucleotide and haematin for growth.

FAMILY MICROCOCCACEAE

The two families so far considered comprise Gram-negative rod-shaped organisms and some of their variants. The family Micrococcaceae in contrast contain spherical cells which are almost invariably Gram-positive. Pigment production is a characteristic of the family. There are six genera, two of which are of interest.

Genus *Micrococcus*

These are aerobic spherical bacteria. One species, *Micrococcus flavus*, is used for the microbiological assay of the antibiotic bacitracin.

Genus *Staphylococcus*

Members of this genus occur widely in nature and are found on the skin and hair of man and animals, they survive drying and can be isolated from dust or from the air, especially in environments where man or animals congregate. Many are pathogenic.

Some members secrete an enzyme called coagulase which has the property of coagulating blood plasma. This property is used to differentiate the two species in the genus. Amongst the biochemical activities of the staphylococci are the production of acids, but not gases, from sugars, and the ability to reduce nitrates to nitrites. They will grow aerobically or anaerobically, thus differing from the micrococci, and are nutritionally exacting. Many staphylococci produce pigments which may vary from white to yellow or orange or red.

Two species are recognised and may be distinguished by two tests.

(1) *Staphylococcus aureus.* Produces acid from mannitol.
Coagulase positive.

(2) *Staphylococcus epidermidis.* Does not produce acid from mannitol.
Coagulase negative.

Staphylococcus aureus is a Gram-positive spherical organism with a diameter of the order of 1·0 μm. The specific name means golden and refers to the colour of the pigment characteristically produced. At one time different species were named according to the colour of the pigment they produced. Pigmentation, because of its variable nature, is not considered a reliable criterion for placing organisms into different species.

Staph. aureus will grow on ordinary culture optimally at 37° C, and will grow in concentrations of sodium chloride of up to 15 per cent. This tolerance to 15 per cent salt is exploited in preparing selective media for isolating *Staph. aureus* from a mixed culture.

The organism is the cause of carbuncles, boils and conjunctivitis, and is pyogenic, that is it tends to produce pus. It can be the cause of more serious conditions such as osteomyelitis and abscesses; some strains can haemolyse red cells due to elaboration of a haemolysin. In addition some strains of *Staph. aureus* can cause food poisoning; these produce a toxin and if foodstuffs in which this toxin is present are eaten symptoms such as diarrhoea and vomiting follow, often within 3 hours and usually not more than 12 hours. It is the difference in time between the ingestion of infected food and the onset of symptoms which is one of the features which distinguish staphylococcal from salmonella food poisoning.

Staphylococcus epidermidis is normally regarded as non-pathogenic. It is found in the same general environmental conditions as *Staph. aureus.* It is less active biochemically and does not produce coagulase. The organism usually produces a white pigment, but as some varieties of *Staph. aureus* produce a white pigment this cannot be taken as a criterion of species.

A Note on the Classification of Micrococci and Staphylococci

It was suggested at the beginning of this chapter that bacterial classification was in a state of constant change. The system adopted in the chapter follows that set out in Bergey's manual. However, much criticism has been levelled against the system adopted in the manual for the classification of the Gram-positive cocci. Recently (1963) a revised scheme has been published; in this scheme it is proposed to divide the genus *Staphylococcus* into six subgroups but not to accord species names to the members of the subgroup. Seven subgroups were recognised amongst the micrococci again without allocation of species names.

FAMILY NEISSERIACEAE

The characteristic feature of members of this family is that they are Gram-negative cocci. The cells occur in pairs or in large aggregates. These organisms are nutritionally exacting. The family contains two genera of which one is of interest in the present context.

Genus *Neisseria*

This genus contains 10 species, two of which are markedly pathogenic, the causal organisms of gonorrhoea and of epidemic cerebrospinal meningitis. *N. gonorrhoea* occurs characteristically as diplococci which are Gram-negative and aerobic, and some growth will occur under anaerobic conditions. The presence of carbon dioxide stimulates growth, and it requires for optimum growth factors present in blood or serum. It is the causal organism of gonorrhoea.

N. meningitidis in general resembles *N. gonorrhoea* but is a strict aerobe. It is the cause of epidemic cerebrospinal meningitis or cerebrospinal fever. The disease is accompanied by acute inflammation of the membranes covering the brain and spinal cord. The disease is endemic in Britain and tends to reach epidemic levels at times when over-crowding and a decline in sanitary standards is forced upon the population. There were two such epidemics associated with the 1914–18 and 1939–45 war periods.

FAMILY LACTOBACILLACEAE

This family contains four genera, all of which contain organisms of interest.

Both rods and coccoid forms are found amongst the Lactobacillaceae. The genera in this family which will be considered are *Diplococcus*, *Streptococcus*, *Leuconostoc* and *Lactobacillus*.

Genus *Diplococcus*

This genus contains one species, *Diplococcus pneumoniae*, commonly called the pneumococcus, one of the causes of lobar pneumonia in man. The organism is a Gram-positive coccus often more ovoid than spherical and occurring characteristically in pairs from which the generic name is derived; it will grow aerobically or anaerobically, optimally at 37° C. One characteristic of the organism is that it is soluble in bile salt solution. The organism is capable of producing a capsule which is responsible for the virulence of the organism and for giving colonies of the organism on solid media a characteristic smooth appearance. If the organism is grown in the laboratory for some time it loses its capsule, giving the non-virulent rough strain. Although *D. pneumoniae* will grow on nutrient broth it grows best if this medium is enriched with blood or serum; 70–80 serological types have been recognised amongst the pneumoccal isolates known.

Genus *Streptococcus*

The streptococci are Gram-positive, spherical or oval organisms often growing in chains. They contain no cytochrome pigments or catalase, are insoluble in bile salts (in contrast to the pneumococcus) and do not produce nitrite from nitrate (in contrast to the staphylococci). All streptococci are nutritionally exacting. They may be aerobic, microaerophilic or anaerobic. To explain aerobic growth with absence of catalase it has been suggested that although hydrogen peroxide is produced during the metabolism of streptococci this potentially toxic metabolite is decomposed by peroxidases.

Bergey's manual lists 17 species and divides them into four groups, the pyogenic (pus producing), viridans (producing green pigment from haemoglobin), enterococcus (living in the intestine) and lactic (producing lactic acid from carbohydrates). The most important criterion, however, for assigning a streptococcal species to one of these four groups is not the characteristic implied by the group name but the range of temperatures over which the species will grow and a number of biochemical differences.

In addition to this grouping, Lancefield characterised the streptococci by serological methods, based on the fact that cellular carbohydrate differs antigenically amongst different strains. In brief the method of serological typing consists of extracting the cells and preparing an antiserum for the extract. This antiserum will now cause a precipitation when it is added to an extract of cells of the same antigenic structure. By this means 15 different groups amongst the pyogenic and enterococcus group have been recognised. They are designated A–Q (I and J are not used). A further nomenclature that has been applied to the streptococci depends on their behaviour towards whole blood. α-Haemolytic streptococci produce a green pigment from haemoglobin. The use of the term viridans, already referred to, derives from this property. β-Haemolytic streptococci lyse red blood cells, producing a clear zone around colonies growing on a solid culture medium containing blood. γ-Haemolytic streptococci are devoid of activity towards blood.

Streptococcus pyogenes. A Gram-positive round to ovoid organism which can grow aerobically and anaerobically and in culture is often seen in chains. The organism is β-haemolytic, belongs to group A of the Lancefield serological classification, produces acids from certain carbohydrates and has an optimum growth temperature of 37° C. It will grow on conventional nutrient media but growth is enhanced if blood or serum is added. The organism is markedly pathogenic to man, producing such conditions as scarlet fever, tonsillitis and acute rheumatism. Amongst the toxins produced are haemolysins, a fibrin-digesting enzyme (streptokinase), hyaluronidase and deoxyribonuclease (streptodornase). The pathogenicity of the organism is, however, attributable to a structural toxic protein. A preparation containing streptokinase and streptodornase is used in medicine to liquefy blood and pus masses in body cavities and is prepared from the medium in which streptococci of Lancefields group C have been growing Group C does not contain streptococci pathogenic to man. *Streptococcus dysgalactiae* and *Streptococcus agalactiae* are two organisms causing mastitis in cattle. *Streptococcus faecalis*, one of the enterococcus group and γ-haemolytic is found in the intestine of man and can cause mild food poisoning. *Streptococcus cremoris*, a member of the lactic group, is a non-pathogenic organism used as a starter in the manufacture of butter and cheese. By conversion of lactose to lactic acid it initiates the physical and chemical changes that result in the production of these two products from milk.

Genus *Leuconostoc*

This genus contains two species which have the property of decomposing sucrose, producing a gummy polyglucose called tran. This has been used as a substitute for human plasma. The organisms concerned are *Leuconostoc mesenteroides* and *Leuconostoc dextranicum.* These organisms are coccoid and Gram-positive and will grow on ordinary culture media. They are often found in sugar refineries where their ability to produce dextran can cause blockages to pipes and valves in the plant. Sucrose is hydrolysed first to glucose and fructose; the fructose is utilised as a source of energy for the growth of the organism while the glucose is polymerised to dextran.

Genus *Lactobacillus*

The lactobacilli occur as long slender rods which are Gram-positive, non-motile and do not form spores; they are usually microaerophilic or anaerobic. The lactobacilli are highly nutritionally exacting. Some members produce lactic acid as the sole end-product when fermenting glucose and are termed homofermentative, while others produce a variety of products and are called heterofermentative. The lactobacilli, because of their nutritionally exacting requirements, are used in the microbiological assay of amino acids and growth factors. *Lb. casei*, *Lb. helveticus*, *Lb. leichmanii* and *Lb. plantarum* have been particularly used in this connection. A description of the assay of riboflavine and nicotinamide may be found in the *B.P.C.*, 1968, pp. 1400 and 1398. *Lb. arabinosus*, stated in the *B.P.C.* as being used for the assay of nictotinamide, is probably the same as *Lb. plantarum*.

FAMILY CORYNEBACTERIACEAE

The corynebacteria are Gram-positive, non-motile rods. They do not form spores or capsules. Staining is often irregular due to the presence in the cytoplasm of metachromatic (volutin) granules. Members of the family are found widely in soil and in animals and plants. There is one genus of interest.

Genus *Corynebacterium*

Bergey's manual lists 32 species of which one is of outstanding importance.

Corynebacterium diphtheriae is the cause of diphtheria, a disease which, thanks to extensive immunisation measures, has largely disappeared.

C. diphtheriae is a slender rod of average dimensions 3 μm $\times$ 0·3 μm. It is non-motile and does not produce spores. It is Gram-positive but well-developed volutin or metachromatic granules in the cytoplasm gives rise to irregular staining. It is aerobic, growing on ordinary media optimally at 37° C. The addition of serum enhances growth. The organism secretes an exotoxin which is related chemically to the cytochrome respiratory pigments; as already stated, the toxin may owe its activity to an interference with protein synthesis. When grown on a medium containing whole blood and potassium tellurite, colonies isolated from cases of the disease fall into three characteristic types and these are associated in turn with a graded degree of severity of the attack. *C. diphtheriae gravis*, from severe cases, gives large grey black colonies with radiating striations. *C. diphtheriae intermedius*, from cases of intermediate severity, gives a small black colony, while *C. diphtheriae mitis*, from mild cases, gives smooth black colonies which, although of the same size as *gravis* colonies these are not striated. The *B.P.* describes vaccines and antitoxins for the prevention and treatment of diphtheria and reagents to detect if subjects have in fact been infected with the organism, or are susceptible to it.

A number of other non-pathogenic species of corynebacteria are often found in the throat. These have been named for convenience diphtheroids.

FAMILY BACILLACEAE

Members of this family are comparatively large, rod-shaped, Gram-positive organisms which produce spores. Many are human and animal pathogens. The family contains two genera, membership of which is determined by their oxygen requirements.

Genus *Bacillus*. Aerobes or facultative anaerobes. Contain catalase.

Genus *Clostridium*. Anaerobic; a few species will tolerate the presence of oxygen. Do not contain catalase.

Genus *Bacillus*

The bacilli are rod-shaped spore-forming organisms which are usually Gram-positive; some are motile, possessing peritrichous flagella. They grow readily in nutrient broth and on nutrient agar. When cultivated on liquid media, organisms often accumulate at the surface, giving rise to a skin or pellicle. Temperature optima and maxima vary widely amongst species in this genus.

Most species are non-pathogenic but one, *B. anthracis*, is the cause of anthrax in man and animals. Some produce antibiotic substances. The bacilli are widely distributed in nature, being found in dust, in the soil and the alimentary tract of man and animals.

B. subtilis is used for the microbiological assay of the antibiotics streptomycin, dihydrostreptomycin, penicillin and phenoxymethylpenicillin, *B. pumilis* for the assay of chlortetracycline, erythromycin, neomycin, oxytetracycline and tetracycline. *B. cereus* is found widely in nature. It is non-pathogenic but some bacteriologists consider *B. anthracis* to be a pathogenic variant of *B. cereus*.

Several species produce polypeptides with antibacterial activity, e.g. tyrothrycin, a mixture of gramicidin and tyrocidine, is produced by *B. brevis*, bacitracin is produced by a strain of *B. subtilis* and by *B. licheniformis* and polymyxin is produced by *B. polymyxa*.

B. stearothermophilus is an example of a thermophilic organism; it grows optimally at 65° C. It produces markedly heat-resistant spores which are used to test the efficiency of sterilisation procedures.

B. anthracis is the only pathogenic member of the genus, producing the disease of anthrax in man and animals. It is a large rod (4–8 μm $\times$ 1–1·5 μm) often growing in chains. The individual cells appear to have square rather than rounded ends, thus their shape approaches more nearly to a true cylinder. It is Gram-positive and non-motile. Under suitable conditions the organism produces a capsule which is composed of a polyglutamic acid. It forms spores which may remain dormant in the soil for many years. Anthrax is predominantly an animal disease but may be contracted by farm workers and others who come into contact with infected animals or animal products.

Genus *Clostridium*

In Bergey's manual are listed 93 species in this genus; 88 are classified as strictly anaerobic while 5 are aerotolerant. The clostridia are spore-forming rods and all the possible positional relationships of spore to cell are found in the genus. The cells are generally Gram-positive and motile with peritrichous flagella. Some are pigmented. They are found widely distributed in nature, in the soil and in the alimentary tract of man and animals. Amongst the clostridia are found virulent human and animal pathogens, organisms that have been used industrially to produce organic solvents such as acetone and butyl alcohol and organisms capable of utilising (fixing) nitrogen gas. The 93 species are differentiated chiefly on spore morphology and biochemical activities.

Nine species of interest are tabulated below (Table 11).

TABLE 11

Pathogens	Solvent producers	Of general interest
Cl. septicum *Cl. perfringens* (*welchii*) *Cl. novyi* (*oedematiens*) *Cl. tetani* *Cl. botulinum*	*Cl. acetobutylicum* *Cl. butylicum*	*Cl. pasteurianum* *Cl. sporogenes*

The clostridia are usually nutritionally exacting and some derive their energy from the fermentation of amino acids.

Pathogenic Clostridia

A serious complication of wounds known as gas-gangrene is caused by various species of clostridia. The species most frequently implicated are *Cl. septicum*, *Cl. perfringens* (often named *Cl. welchii*) and *Cl. novyi* (or *Cl. oedematiens*). A variety of tissue-destroying enzymes are produced and in some cases gaseous products from the anaerobic breakdown of glycogen and of proteins.

Cl. septicum, a bacillus 3–10 μm × 0·6 μm, which is motile by peritrichous flagella, producing an oval central or subterminal spore larger than the original rod from which it arises.

Cl. perfringens (*welchii*), a bacillus 4–6 μm × 1 μm, non-motile, the spores are oval central or subterminal but are not larger than the vegetative rod.

BIOCHEMICAL ACTIVITIES OF SOME CLOSTRIDIA

	Glucose	Lactose	Sucrose	Maltose	Salicin
Cl. septicum	+	+	—	+	+
Cl. perfringens	+	+	+	+	v
Cl. novyi	+	—	—	+	—
Cl. tetani	—	—	—	—	—
Cl. botulinum	+	v	v	+	v
Cl. pasteurianum	+	—	+	+	—
Cl. sporogenes	+	—	—	+	v

+ fermented; — not fermented; v variable results obtained amongst different isolations.

Cl. novyi (*Cl. oedematiens*), a bacillus 2·5–5·0 μm × 0·8 μm, motile by peritrichous flagella, producing ovoid spores usually subterminal and slightly larger than the vegetative rod.

All three species are obligate anaerobes, and ferment carbohydrates. A summary of carbohydrate fermentation as a means of differentiating six clostridial species is given above.

Cl. tetani, the causal organism of tetanus or lock-jaw, is 2–5 μm × 0·5 μm. The spores are spherical, terminal and wider than the bacillus, giving rise to the so-called drum-stick appearance. This organism is without action on carbohydrates.

Cl. botulinum is a rod 4–6 μm × 0·9–1·2 μm. The spores are subterminal oval and slightly larger than the vegetative form. The organism is motile with peritrichous flagella. This organism is the cause of botulism, an invariably fatal condition caused by ingestion of an exotoxin produced by the organism. Poisoning usually occurs after eating food in which the organism has been able to grow.

Cl. pasteurianum is a non-pathogenic species found in the soil. It is able to utilise or fix atmospheric nitrogen and was the first nitrogen-fixing organism to be recognised.

Cl. sporogenes, a non-pathogenic species found in the soil and in the alimentary tract of man and animals. It is used as a control organism for anaerobic media in sterility tests.

Another biochemical division that has been made amongst this group of organisms is to designate them saccharolytic as measured by vigorous acid and gas production from carbohydrates; proteolytic as measured by the ability to digest protein, producing unpleasant smelling, volatile products; and purinolytic as measured by their ability to decompose purines. *Cl. septicum*, *Cl. perfringens* and the organic solvent-producing clostridia are examples of saccharolytic organisms; *Cl. tetani* and *Cl. sporogenes* are examples of proteolytic organisms. *Cl. acidiurici* is a purinolytic organism, decomposing uric acid to carbon dioxide, ammonia, acetate and formate. *Cl. butylicum* and *Cl. acetobutylicum* produce acetone and butanol from glucose and also from raw materials such as molasses, starch, potato mash and in the case of the latter organism, maize mash. This reaction has been and in some countries still is, exploited to produce the important solvents acetone and butanol. Both these organisms can fix atmospheric nitrogen.

ORDER ACTINOMYCETALES

This order is characterised by the fact that its members often grow in a mycelial form with branching reminiscent of the fungi; other criteria, however, lead unequivocally to the view that the actinomycetes are rightly classified with the bacteria.

There are four families in the order, three of which contain genera of interest.

Family	Genus
Mycobacteriaceae	*Mycobacterium*
Streptomycetaceae	*Streptomyces*
Nocardiaceae	*Nocardia*

FAMILY MYCOBACTERIACEAE

Genus *Mycobacterium*

The mycobacteria show, amongst the *Actinomycetales*, the least tendency to branch. They are usually seen as slender, straight or curved rods. They do not form spores or capsules and are non-motile and aerobic. They stain with difficulty but once stained, for example by heating with the stain, are then difficult to destain even with acid. This property, called acid fastness, distinguishes them from all other vegetative bacteria.

Mycobacterium tuberculosis. This is the causal organism of tuberculosis in man. Allied varieties cause tuberculosis in cattle and various lesions in birds and rodents. The organism occurs as thin, straight or slightly curved rods, 3·0 μm $\times$ 0·3 μm, which are acid fast, and may stain evenly or unevenly, giving the rod the appearance of a short string of beads. Old cultures may show branching. It is aerobic, non-capsulated and non-motile and does not form spores, although it will survive desiccation for several months; this may be because of the high wax content of the cells (in the human variety this may be as much as 23 per cent). The cellular wax may inhibit water loss and fatal desiccation.

The optimum growth temperature is 37° C. When first isolated *Myco. tuberculosis* cannot be grown on the usual laboratory medium but needs media enriched with egg yolk, blood or serum. Thereafter growth will occur on an ordinary medium provided glycerol is added, but 10 days may elapse before growth becomes visible.

Mycobacterium tuberculosis var. *bovis.* This organism is the cause of a usually fatal pulmonary tuberculosis in cattle. It is found in some 10 per cent of cases in humans dying of tuberculosis and has been isolated from about a quarter of all cases of tuberculosis in children. A campaign to eradicate infected cattle from dairy herds is being actively pursued, as infection can be spread by drinking contaminated milk. A strain of this variety, first isolated in 1921 by Calmette and Guérin, is used to prepare a vaccine (B.C.G. vaccine) which is described in the *British Pharmacopoeia*, and used for conferring immunity against tuberculosis. Therein are also described products called tuberculins prepared from the culture filtrate in which *Myco. tuberculosis* has been grown (both human and bovine types are used). These products are used as diagnostic reagents to determine if a human subject has at any time been infected with *Myco. tuberculosis.* A positive tuberculin test is not necessarily an indication of clinical tuberculosis (Volume 5).

Mycobacterium leprae has been implicated as the causal organism of leprosy.

Mycobacterium phlei is a non-pathogenic species found widely in nature.

Mycobacterium butyricum, another non-pathogen, is one of the causes of rancidity in butter.

FAMILY STREPTOMYCETACEAE

Genus *Streptomyces*

The streptomycetes grow in the form of a multibranched mycelium, thus outwardly resembling the growth pattern of many fungi. They are anaerobic or microaerophilic. Spores or conidia are produced in chains; these spores are reproductive bodies and thus differ in function from the spores of the other bacterial orders so far described. The process of their formation, however, is different from the process of spore formation in the true fungi. The streptomycetes are found widely in the soil, and from members of this genus have come the majority of antibiotics used in medicine (Table 12).

The streptomycetes may be grown on ordinary laboratory media and cultures very often have a characteristic earthy smell.

TABLE 12

SOME ANTIBIOTICS PRODUCED BY STREPTOMYCETES

Streptomyces species	Antibiotic
S. griseus	Streptomycin
S. venezuelae	Chloramphenicol
S. aureofaciens	Chlortetracycline
S. rimosus	Oxytetracycline
S. viridifaciens	Tetracycline
S. erythreus	Erythromycin
S. fradiae	Neomycin
S. noursei	Nystatin
S. floridae, *S. puniceus*	Viomycin
S. niveus, *S. spheroides*	Novobiocin
S. nodosus	Amphotericin B
S. orchidaceus, *S. garyphalus*	Cycloserine
S. halstedii	Carbcmycin
S. ambofaciens	Spiramycin

Vitamin B_{12} (cyanocobalamin) is also produced commercially from *S. griseus*.

Because of their importance in antibiotic production much study has been devoted to methods of culture with a view to enhanced yields of antibiotic.

FAMILY NOCARDIACEAE

Genus *Nocardia*

In this genus the mycelium tends to fragment easily into bocillary elements, and unlike the streptomycetes its members are aerobic.

Nocardias occur in the soil and are mostly non-pathogenic. A few can cause disease in man and animals. Thus *N. farcinica* infect the lungs and lymphatic system in cattle causing the disease known as farcy.

The antibiotics ristocetin is produced from *Nocardia lurida*.

ORDER: SPIROCHAETALES

This order is divided into two families each consisting of three genera.

The Spirochaetales occur as thin spiral organisms which may be as great as 500 μm in length. They are motile but not by means of flagella. They multiply by transverse fission.

Family	Genus
Spirochaetaceae	*Spirochaeta* *Saprospira* *Cristispira*
Treponemataceae	*Borrelia* *Treponema* *Leptospira*

Of the six genera listed it is those in the family Treponemataceae which contain human pathogens.

Genus *Borrelia*

In this group are found some 14 species which cause relapsing fever in man; the most commonly encountered in Europe is *B. recurrentis.*

Vincent's angina, characterised by lesions in the mouth, is caused by *B. vincenti.*

Genus *Treponema*

Treponema pallidum is the causal organism of syphilis. It is a thin spiral cell of average length 10 μm and is just over 0·1 μm in diameter. There is no unequivocal evidence that it has been cultivated outside a living animal. It will retain its viability in blood or serum drawn from a syphilitic subject for 3–4 days even at 4° C. It is thus a disease which may be transmitted by blood transfusion.

Because of its social consequence there are well-established tests for the diagnosis of the disease.

Genus *Leptospira*

Leptospira icterohaemorrhagiae is the cause of a type of jaundice in man known as Weil's disease. It has the same general dimensions as *Tr. pallidum.* This organism may be grown in laboratory culture medium containing serum.

RICKETTSIAE AND VIRUSES

The Rickettsiae and viruses are the smallest structures which appear to be endowed with life, in so far that they are able to reproduce. They are entirely parasitic, living and reproducing in the cells of other living organisms, ranging from plants, including the bacteria, to man and causing a variety of diseases. The study of viruses has become the province of the specialist and virology a science in its own right.

THE RICKETTSIAE

A scheme of classification of the order Rickettsiales is given in Bergey's manual; only one genus will be considered which contains three species of interest.

Genus *Rickettsia*

Rickettsiae are found as small, rod-shaped or coccoid cells, 0·3 μm × 0·25 μm. They can be stained by special procedures. Electron microscopy has suggested the presence of a capsule, a cell wall and a nucleus. Division is by binary fission. Rickettsiae may be cultivated in the blood of laboratory animals or in the yolk sac of the embryo of the domestic fowl and it is by this method that the organism is grown to produce vaccines.

Infection with rickettsiae give rise to a variety of typhus infections, the intermediate carriers by which man is infected being lice, fleas, ticks or mites, where rickettsiae can occur without harm to these arthropod hosts.

Amongst the diseases caused by rickettsiae are epidemic typhus, trench fever and murine typhus, caused by *R. prowazekii*, *R. quintana* and *R. typhi* (Fig. 26), respectively.

A vaccine is described in the *British Pharmacopoeia* which is prepared from the rickettsiae of epidemic and murine typhus.

THE VIRUSES

The viruses, because of their small size, escaped detection until comparatively recently as the cause of certain infective diseases of animals and plants (for example foot and mouth disease and tobacco mosaic), and in fact it was not until 1930 and by physicochemical techniques that the existence of the infective agent or virus, hitherto inferred, was demonstrated. A virus was 'crystallised' in 1935 and the advent of electron microscopy made possible the photographing of viruses. The existence of an invisible infective agent, however, was suggested by Louis Pasteur in 1892 as the cause of rabies in his classic study of this disease.

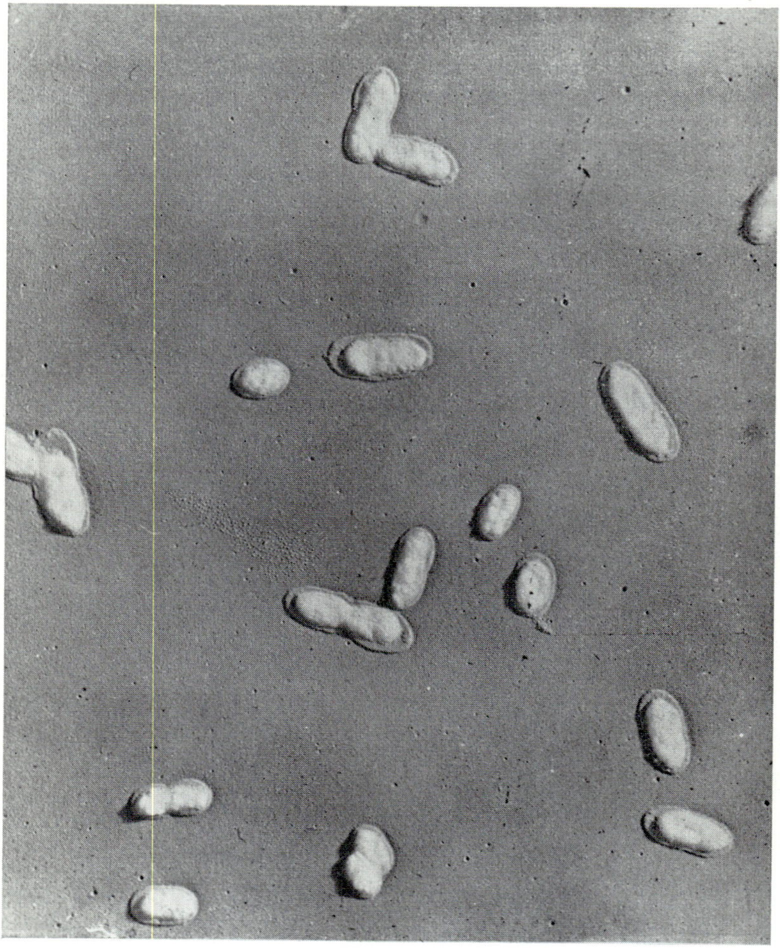

FIG. 26. *Rickettsia typhi.* The causal organism of murine typhus. Electron micrograph, ×36,000. (Parke, Davis and Co., England)

Where it has been possible to prepare virus material free from host cells, techniques more familiar to the physical chemist than to the microbiologist have been used to determine the structure of many viruses. Likewise analyses of viruses have given insight into their chemical composition. The infective virus particle is called the viron.

The Chemical Nature of Viruses

Three types of macromolecule have been found in all the viruses so far, namely, ribonucleic acid (RNA), deoxyribonucleic acid (DNA) and protein, and occur in the virus as an RNA-protein complex or a DNA-protein complex. Viruses, therefore, are composed of nucleoprotein. More detailed studies have shown that the protein exists on the outside of the particle and is known as the capsid. The nucleic acid component is enclosed within the capsid. Muramic acid has recently been demonstrated as a constituent of some of the larger viruses, suggesting the presence of a pepticloglycan as found in the cell wall of bacteria. This finding is in accord with the known susceptibility of the larger viruses to penicillin.

The Shape and Size of Viruses

The viruses usually exist as spherical, cubical or rod-shaped particles. The largest may be 350–500 *n*m in diameter, the smallest 10 *nm*. It will be recalled that the rod-shaped micro-organism *Serratia marcescens* has an average length of 750 *n*m. Viruses which attack bacteria (bacteriophages) have polyhedral heads with hexagonal faces and from the head a short tube-like structure emerges.

The size of virus particles has been determined by the use of special filters made of collodion and of known pore size. By filtering solutions containing virus particles and from a knowledge of the average pore size of the filter just preventing the particles from passing an approximation of the size of the virus can be made. This technique was much used in early work on viruses.

Methods based on rate of sedimentation in an ultracentrifuge and direct measurement in the electron microscope have largely replaced filtration methods. Electron microscopy also gives some indication of the shape of viruses, and this combined with other techniques has shown that individual viruses possess an incredible individualistic architecture as shown in the virus models in Fig. 27, and is due to the individual shape assumed by the protein units of the capsid which are called the capsomeres.

Biological Properties of Viruses

The cardinal feature of biological activity amongst the viruses, including the viruses which attack bacteria (the bacteriophages), is

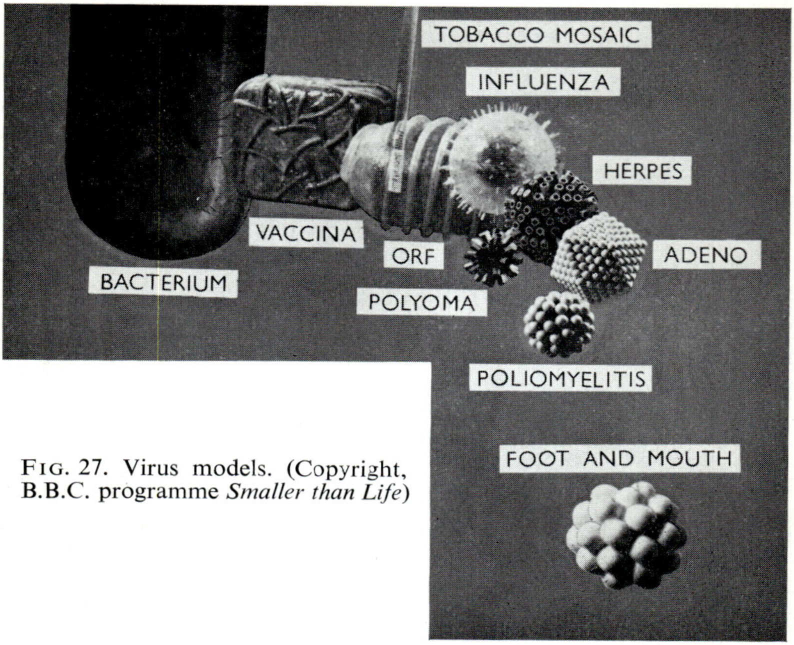

FIG. 27. Virus models. (Copyright, B.B.C. programme *Smaller than Life*)

their ability to replicate themselves in living tissue. When they invade living cells they seem able to redirect the biosynthetic potential of that cell to produce more virus substance. Outside the cell they show no metabolic activity, as for example uptake of oxygen or utilisation of externally supplied foodstuffs such as glucose or amino acids.

Following the infection of a susceptible living organism by a virus, the virus particle disintegrates. The nucleic acid is released and is thought to be responsible for the reorganisation of the metabolism of the host cell towards the synthesis of new virus.

A feature of some virus infections is the appearance in the host cell of masses varying from 1 to 30 μm in diameter, and called inclusion bodies. These have been shown in many cases to consist of a mass of virus particles surrounded by a membrane. These bodies possess characteristic staining properties and when present their size and position in the cell may be of value in diagnosing the nature of the infection.

Virus infection can give rise in the human or animal host to specific antibodies and a single infection from many viral diseases usually confers immunity for life. The common cold and influenza virus infections are notable exceptions to this rule.

Reaction of Viruses to the Physical Environment

Viruses are usually destroyed by moderate heating, 50° C, for half an hour but will survive low temperatures, for example −60° C.

The majority of viruses do not survive for more than about 6 hours outside the body but this period is long enough to enable cross infection from virus-carrying particles to occur and this is the normal method of spread of infection in the case of the common cold and influenza.

The viruses causing smallpox and poliomyelitis, however, survive outside the body for much longer periods.

Viruses are best destroyed by disinfectants which are oxidising agents, as for example iodine and the hypochlorites.

Virus Diseases

Amongst the more important diseases of man attributable to viruses are those shown in the table on page 148.

Diseases of animals which are of viral origin include swine fever, foot and mouth disease, canine distemper, hard pad, cowpox, rabies, psittacosis and fowl pest. Psittacosis, rabies and cowpox are also diseases of man.

Virus diseases of plants include tobacco mosaic virus and leaf curl in potato. Many other observed changes in plant appearance, such as warty or curled leaves or patches of discoloration on leaves, are often indicative of viral infection.

The Cultivation of Viruses

It is a prerequisite of viral growth that it occurs in living tissue. This may be achieved by growing the virus in a living organism or in isolated but living tissue. The virus used to prepare smallpox vaccine is grown on the skin of sheep or calves, or in a tissue much used in virology, the living embryo of the domestic fowl. The latter tissue is also used to grow the yellow fever virus for the preparation of the appropriate vaccine.

The fowl embryo presents a living tissue which is likely to be sterile and on which many viruses grow in a characteristic manner;

Disease	Tissue or organ system affected
Chicken-pox	Skin
Common cold	Respiratory system
Cow-pox	Mucous membrane and skin
Croup	Respiratory system
German measles	General infection
Homologous serum jaundice	Liver
Infective hepatitis	Liver
Influenza*	Respiratory system
Measles*	General infection
Mumps	Parotid salivary glands
Poliomyelitis*	Central nervous system
Pneumonia (one cause)	Respiratory system
Rabies*	Central nervous system
Smallpox*	General infection
Warts	Skin
Yellow fever*	General infection

* Vaccines for protection from these diseases are described in the *British Pharmacopoeia*, 1968.

this tissue will not produce antibodies against the inoculated virus.

The experimental technique consists of opening the egg 12–15 days after fertilisation and inoculating the vaccine on to the chorioallantoic membrane; after three days the virus, if the inoculation has been successful, will have grown on the membranes producing characteristic lesions. This viral culture may be used for fundamental studies of viruses or used to prepare viral vaccines.

For the preparation of poliomyelitis vaccine a culture of the kidney tissue of monkeys, usually the Indian rhesus, is used. The kidneys are removed from the animal, using strict aseptic precautions, ground and suspended in a complex medium in which kidney cells will remain alive and continue to metabolise. This forms the substrate for the growth of the poliomyelitis virus.

Interferon

It has been known for some time that if man or an animal is suffering a virus infection, an infection by a second virus may in some cases be impossible. The first virus is said to interfere with the infective activity of the second. Recently a specific substance responsible for the effect and called interferon has been isolated and promises to be of use in the treatment of virus disease.

BACTERIOPHAGES

The viruses attacking bacteria are known by the name of bacteriophages or phages. Structurally they consist of a polyhedral head with hexagonal faces from which emerges a short tube, often called a tail. If a bacteriophage invades a culture of susceptible bacteria the following sequence of events is known to occur. The bacteriophage particles become attached to the bacterial surface by means of the tube or tail. Specificity of bacteriophage for host is thought to reside in the shape of the end of this tube which is able to achieve a fit with some receptor site of complementary shape at the bacterial surface. Subsequent to absorption a lytic enzyme in the tail destroys the cell surface directly beneath the point of contact and the deoxyribosenucleic acid contained within the head passes into the interior of the bacterial cell. A similar event now occurs, as happens when a plant or animal cell is invaded by a virus. The DNA from the bacteriophage takes over from the DNA of the bacterial nucleus and starts to dictate the biosynthetic processes of the metabolites available within the bacterial cell, resulting in the production of new bacteriophage particles.

Finally the bacterial cell bursts liberating new bacteriophage particles. For each single phage particle originally infecting a cell some 200–400 new particles are produced.

The manifestation of bacteriophage activity in a normal turbid bacterial culture is the clearing of the culture due to the lysis of the organisms. The released phage particles are too small to confer visible turbidity.

Because of the high degree of specificity of bacteriophage for host, bacteriophages of known specificity are used in the classification of bacteria, by measuring the incidence of lysis.

THE FUNGI

The study of this branch of the plant kingdom, the fungi, has been the domain of the specialist for over a century and the science is known as mycology.

Fungi are of importance in pharmacy in the following three connections:

1. They may be a source of antibiotics, and such chemicals

as citric, gluconic and fumaric acids, and are used in the preparation of steroid intermediates.

2. They may cause spoilage or contamination of pharmaceutical products.

3. Some fungi are pathogenic and it may be necessary to evaluate potential antifungal agents; thus a knowledge of the cultivation of pathogenic fungi is of importance.

Growth, Form and Classification

The fungi may be unicellular, as exemplified by many yeasts; the vast majority, however, grow in the form of multibranched filaments or threads called hyphae. These threads or tubes may or may not bear cross walls or septa. Reproduction is usually carried out by the production of spores. These may be produced by a sexual process (perfect stage) but more commonly spores are produced asexually (imperfect stage). Most yeasts reproduce by budding, although reproduction by fission and by sexual and asexual spore production can occur. Chlorophyll is characteristically absent in the fungi.

For classification, the true fungi have been divided first into the lower fungi and the higher fungi. The lower fungi contains one class, the Phycomycetes; the higher fungi three classes, the Ascomycetes, Basidiomycetes and the Fungi imperfecti.

Phycomycetes

This class contains some unicellular members. Those that grow in the mycelial form have aseptate hyphae.

Ascomycetes

The spores in the sexual or perfect stage are borne in a specialised organ called an ascus. Hyphae, when present, are septate. This class includes the yeasts.

Basidiomycetes

The spores of the sexual stage are borne on specialised cells called basidia. The mycelia are septate. The smuts, rusts and the familiar mushrooms and toadstools belong to this class.

Fungi Imperfecti

This group is characterised by possessing no sexual stage, only the asexual or imperfect stage. Important members of this class include the aspergilli and penicillia.

Although the word fungus or, used collectively, the fungi, refers to all the groups of organisms mentioned above, common usage tends to give a breakdown in terminology as follows:

Yeasts: The yeasts.
Moulds: Phycomycetes, Ascomycetes (other than yeasts), Fungi imperfecti.
Fungi: Basidiomycetes—more especially the familiar mushrooms, toadstools and bracket fungi.

In examining pharmaceutical material it is from the yeasts and moulds that most if not all contaminants will be found.

The Growth of Fungi

Most fungi encountered as contaminants in pharmaceutical products will grow on media similar to that used to grow bacteria. Growth is favoured, however, if the proportion of carbohydrate in the medium is increased in relation to the proportion of nitrogenous constituents.

Thus, media for the cultivation of fungi often contains additional glucose, malt, sucrose or wort. The optimum pH for mould growth is usually on the acid side of neutrality. The pH of culture media for the growth of moulds is usually 5–6; this, while being entirely suitable for most common moulds, at the same time discourages bacterial growth and thus renders the medium selective.

The optimum temperature varies widely from species to species but in general the common moulds will grow better at 22–25° C than most human pathogenic and commensal bacteria. It is customary, therefore, to incubate mould cultures at lower temperatures than bacterial cultures.

The *United States Pharmacopoeia* directs that the presence of fungi should be tested for in certain pharmaceutical products on a liquid medium incubated at 25° C. This temperature, however, is likely to allow the growth of bacteria which do not grow well at 37° C but grow well at lower temperatures, so that it should not be assumed that the incubation temperature provides a selective means of detecting fungi and bacteria.

The Chemical Structure of Fungi

In the majority of fungi the cell wall consists of a polymer of *N*-acetylglucosamine called chitin, a substance found also in the exoskeleton of crustacea and insects. Some fungal cell walls are composed of cellulose. Within the cell wall is a nucleated cytoplasm. The usual complement of carbohydrates, lipids, nucleic acids and proteins including enzymes have been isolated from the cytoplasm.

Metabolism. The fungi are heterotrophic and carry out a range of metabolic activities as diversified as found amongst the bacteria.

Commercially the production of alcoholic beverages (brewing), the production of penicillin and griseofulvin, steroid intermediates, citric acid and certain enzymes used in commerce are the results of the biochemical activities of fungi (yeasts and moulds).

Detailed studies amongst the fungi show that their metabolic activity follows very similar patterns to those in bacteria. Thus the Embden–Meyerhof, Entner–Doudoroff, hexose monophosphate and pentose pathways, the direct oxidation pathway and the citric acid cycle have been found in the fungi. The sequence for terminal oxidation follows the same patterns as in the bacteria. Transamination, nitrate reduction, breakdown of aromatic rings, nucleic acid and protein metabolism, to mention only a few examples, follow similar pathways in the bacteria and fungi.

FUNGI OF PHARMACEUTICAL IMPORTANCE

These may usefully be divided into three categories; those producing useful chemicals, those used in the partial synthesis of medicinal chemicals and those which produce disease.

Fungi Producing Medicinal and Horticultural Chemicals

Antibiotics. The number of useful antibiotics produced by the fungi is much less than those produced by the streptomycetes (page 141); those of importance are shown in Table 13. Fig. 23 illustrates a specimen of the organism from which penicillin was first obtained.

TABLE 13

ANTIBIOTICS FROM FUNGI

Fungal species	Antibiotic
Penicillium chrysogenum	Penicillin
Penicillium urticae	Griseofulvin
Penicillium raistrickii	Griseofulvin
Aspergillus fumigatus	Fumagillin

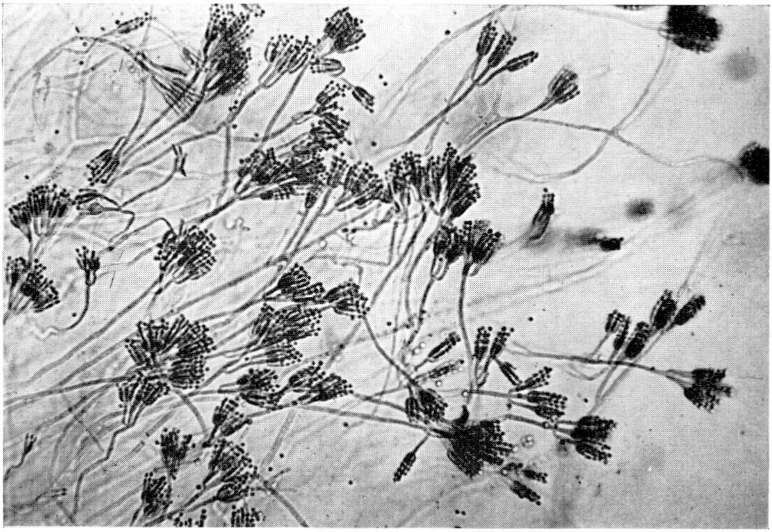

FIG. 23. *Penicillium notatum.* The organism from which penicillin was originally produced. (J. and E. Sturge Ltd)

Citric acid. This is prepared from molasses using *Aspergillus niger* (Fig. 24). As some strains of this organism also produce oxalic acid from molasses it is essential to select a strain of *A. niger* which produces only citric acid. Citric acid production has been achieved both by deep and surface culture methods. Gibberellic acid produced by *Gibberella fujikuroi* may be used in horticulture as a growth hormone, and several interesting possibilities as to its use have been promulgated.

Fungi Used in Synthesis of Medicinal Chemical Intermediates

Synthesis of steroids. One stage in the manufacture of cortisone and hydrocortisone from readily available steroids involves an oxidation process at position 11 in the steroid ring system, that

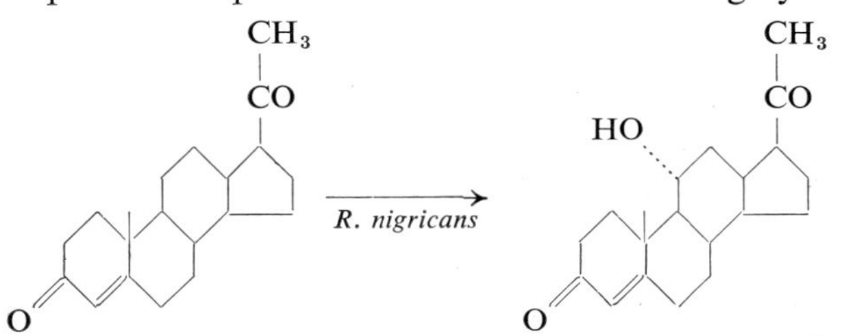

Progesterone 11,α-Hydroxyprogesterone

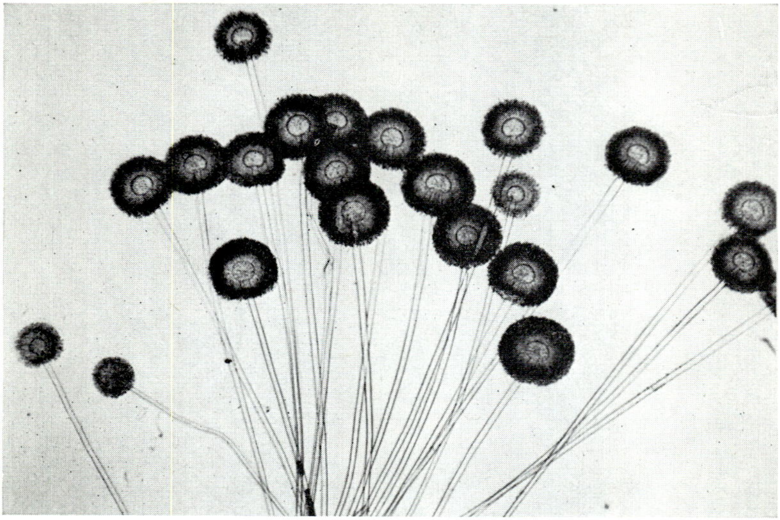

FIG. 24. *Aspergillus niger*. Citric acid producing strain.
(J. and E. Sturge Ltd)

is the conversion of progesterone to 11,α-hydroxyprogesterone. This transformation, difficult to achieve by chemical methods, is performed in high yield by the action of the fungus *Rhizopus nigricans*. This and many other examples of the biological oxidations of steroids are now well substantiated and used in the manufacture of substances of importance in medicine.

Pathogenic Fungi

Fungi can cause a variety of diseases in man. Amongst those commonly met with in this country are the following:

Thrush (moniliasis). This condition is usually met with as a superficial infection of the skin and mucous membranes. It is commonly found as an infection of the mucous membranes of the mouth of infants. It is caused by *Candida albicans*, one of the Fungi imperfecti.

Ringworm (tinea). Ringworm is an infection of the skin caused by several fungi called for convenience the dermatophytes. Three genera, all belonging to the class Fungi imperfecti, are implicated in ringworm infections; they are *Epidermophyton*, *Microsporum* and *Trichophyton*.

References

Bergey's Manual of Determinative Bacteriology. 7th Ed. by Breed, R. S. Murray, E. G. D., and Smith, N. R. Baillière, Tindall and Cox, London, 1957

Topley and Wilson's *Principles of Bacteriology and Immunity*, Vols. 1 and 2. 5th Ed. by Wilson, G. S., and Miles, A. A. Arnold, London, 1964

Mackie and McCartney's *Handbook of Bacteriology*. 10th Ed. by Cruickshank, R. Livingstone, Edinburgh, 1962

An Introduction to Industrial Mycology. 5th Ed. Smith, G. Arnold, London, 1960

Physiology of the Fungi. Cochrane, V. W. Chapman and Hall, London, 1958

An Introduction to the Biology of Yeasts. Ingram, M. Pitman, London, 1955

The Viruses. Vols. I to III. Ed. Burnet, F. M., and Stanley, W. M. Academic Press, New York, 1959

Textbook of Virology. Rhodes, A. J., and van Rooyen, C. E. Williams and Wilkins, Baltimore, 1962

Microbial Classification. 12th Symposium, Soc. gen. Microbiol. Ed. Ainsworth, G. C., and Sneath, P. H. A. Cambridge University Press, 1962

A classification of micrococci and staphylococci based on physiological and biochemical tests. Baird-Parker, A. C. *J. gen. Microbiol.*, 1963, **30,** 409

Manual for the Identification of Medical Bacteria. Cowan, S. T. and Steel, K. J. Cambridge University Press, 1965

Our current understanding of viruses. Wildy, P. *The Practitioner*, 1967, 199, 597

The biology of viruses. Smith, K. M. Oxford University Press, 1965.

Organism and Control in Prokaryctic and Eukaryotic Cells. 20th Symposium, Soc. gen. Microbiol. Ed., Charles, H. P. and Knight, B. C. J. G. Cambridge University Press, 1970.

Taxonomic significance of fermentative versus oxidative metabolism of carbohydrates by various gram-negative bacteria. Hugh, R., and Leifson, E. *J. Bact.*, 1953, **66,** 24

The Literature of Microbiology

Chemical Abstracts devotes a section to microbiology, concentrating mainly on the chemical aspects of the subject. The decennial and annual indices of this work form a useful source of material.

Review Publications. These give comprehensive reviews on special aspects of microbiology. The following are of importance:

Advances in Applied Microbiology
Advances in Microbial Physiology
Annual Review of Microbiology
Bacteriological Reviews
Progress in Industrial Microbiology

Journals. These publish results of research and occasionally reviews in microbiology. The biochemical journals print aspects of microbiological chemistry. The more important journals are listed below:

Acta Pathologica et Microbiologica Scandinavica
Annales de l'Institute Pasteur
Applied Microbiology
Archives of Biochemistry
Biochemical Journal

Biochemica et Biophysica Acta
British Journal of Experimental Pathology
Journal of Applied Bacteriology
Journal of Bacteriology
Journal of Biological Chemistry
Journal of General Microbiology
Journal of Hygiene
Journal of Pathology and Bacteriology

INDEX

Acetone-butanol fermentation, 95–96
Acetylmethylcarbinol formation, 96–7
Acid-fast organisms, staining, 30–1
Actinomycetales, general properties, 139–41
Adenine, 46, 87
ADP, 87, 92, 94
Aerobacter, general properties, 123
— *aerogenes*, 2
— — capsule, 17
— — cultivation, 49
— — identification, 97, 125–7
Agglutin absorption test, 27–8
Agglutination, 27–8
Alanine, 45, 46
— and cell walls, 6–10
— limitation and sporulation in *B. mycoides*, 19
Amino acid decarboxylases, 107–8
— acids, bacterial metabolism of, 107–8
— — — utilisation of, 44
p-Aminobenzoic acid, 46, 47
Amphotericin, 141
Anaerobes, cultivation and growth, 51–4
Aneurine hydrochloride, 47
Antibiotics, 70
— produced by fungi, 152
— — by *Streptomyces*, 141
Antibody, 27
Antigens, bacteria as, 27–8
Antimetabolites, bacterial, 68–9
Antiseptics, 70
Arginine, 45, 46
Aromatic rings, bacterial metabolism of, 112–13
Ascomycetes, 150
Asparagine, 46
Aspartic acid, 45, 46
Aspergillus fumigatus, antibiotics from, 152
Aspergillus niger, citric acid from, 153
ATP, 87, 90–1, 94

Bacillaceae, general properties, 135–138
Bacillus, general properties, 135–6
— spore, 18
— *anthracis*, capsule, 17
— — effect of cold on, 63
— — heat sterilisation of, 61
— — nutritional requirements, 46
— — toxin, 26
— *polymyxa*, spores, 20
— *subtilis*, sporulation, 20
— — teichoic acid in cell walls, 8–9
— — transamination, 105
Bacteria, anatomy, 3–23
— as antigens, 27–8
— autotrophic, 48
— cell wall, 3–10
— — isolation of, 4
— continuous culture, 42
— cultivation of, 37–8, 48–57
— cytoplasm, 13–14
— cytoplasmic membrane, 10–13
— — — separation of, 10–11
— effect of desiccation, 63
— enzymes, 25
— — and enzyme action, 72–81
— extramural layers, 16–18
— — fimbriae, 16
— flagella, 16
— genetic exchange in, 33–6
— growth, chemical inhibition of, 67–70
— — effect of environmental factors on, 43–70
— — — of irradiation on, 65–7
— — — of osmotic pressure on, 63–5
— — — of pH on, 57–9
— — — of temperature on, 59–62
— — mean generation time, 42–3
— — measurement of, 38

Bacteria, growth, synchronous, 40–1
— heterotrophic, 49–50
— — nutritional requirements, 43–57
— identification of, 119
— luminosity, 24
— metabolism, anaerobic reactions, 85–6
— — cytochrome-linked dehydrogenases, 84–5, 89
— — inhibitors of, 67–70, 115
— — NAD-linked dehydrogenases, 84
— — NADP-linked dehydrogenases, 84
— — of amino acids, 103–8
— — of aromatic rings, 112–13
— — of fats, 113–14
— — of nitrogen-containing compounds, 101–12
— — of nucleic acids, 108–11
— — of proteins, 101–3
— — oxidation reduction potentials, 54–6, 83–5
— — oxidative processes in, 82–91
— — Pasteur effect, 87
— morphology, 1–3
— nutritional classification of, 48–50
— oxygen requirements of, 50–7
— pigments, 23
— reproduction, 31–6
— shape, 1–3
— spore, 18–23
— staining of, 28–31
— toxins, 24–7
Bacteriophage, 149
Basidiomycetes, 150
Bergey order numbers, 119
Biotin, 46, 47
Bismuth sulphite agar, 126
Bordetella, general properties, 128
Borrelia, general properties, 142
Brewer's medium, 53
Brucella, general properties, 128
Brucellaceae, general properties, 127–9

Candida albicans, 154
Capsules, 17–18, 27
Carbomycin, 141
Carbon, utilisation by bacteria, 43–4
Carboxylic acid group, and amino acid metabolism, 107–8
Catalase, 86, 89
Chemosynthetic autotrophic bacteria, 48
Chick–Martin test, 124
Chloramphenicol, 141
Chloratetracycline, 141
Citric acid, produced from fungi, 153
Classification of micro-organisms, 117–19
Clostridium, biochemical activities of, 137
— degradation of nucleic acids, 111
— general properties, 136–8
— nutritional requirements, 46
— pathogenic, 137–8
— spore, 18
— *botulinum*, toxin, 25–6
— *novyi*, toxin, 26
— *oedematiens*, lecithinase, production, 25
— *perfingens*, enzymes, 25
— — (*welchii*), 2, 3
— — — toxin, 26
— *septicum*, toxin, 26
— *sporogenes*, Stickland reaction, 106
— *tetani*, toxin, 26
Coagulase, 26, 129
Coenzymes, bacterial, 73–6
Cold, effect on bacteria, 62–3
Conjugation, bacterial, 33–5
Continuous culture, 42
Corynebacteriaceae, general properties, 134
Corynebacterium, general properties, 134–5
— *diphtheriae*, 2
— — nutritional requirements, 46
— — toxin, 26
Cultivation of bacteria, 37–8, 48–57
— — anaerobic, 51–7
— — non-exacting, 49–50
Cycloserine, 141
Cysteine, 45

Cystine, 22, 44, 45
Cytochromes, 84–5
Cytoplasm, bacterial, 13–15
Cytoplasmic membrane, bacterial, 10–13

Decanal, 24
Decimal reduction time, 62
Dehydrogenation, 82
Deoxyribose, 108
Desiccation, effect on bacteria, 63
Dilution rate, 42
Diplococcus, general properties, 132
— *pneumoniae*, 2, 132
— — capsule, 17
— — nutritional requirements, 46
Disinfectants, 73
DNA, bacterial, effect of irradiation on, 65–7
— — structure, 108–11
Dodecanal, 24

Embden–Meyerhof glycolysis cycle, 91–4
Enterobacteriaceae, general properties, 121–7
— laboratory differentiation of, 125–127
Environment, effect on bacterial growth, 43–8
Enzyme action, bacterial effect of pH on, 77
— — — inhibitors on, 79–80, 115
— — — substrate concentration on, 77–8
— — — temperature on, 79
Enzymes, bacterial, 72–81
Epidermophyton, 154
Erythromycin, 141
Escherichia, general properties, 121
— *coli*, capsule, 18
— — cell wall, 9–10
— — cultivation, 49, 125–7
— — effect of bismuth sulphite on, 126
— — — of temperature on growth, 58
— — — of pH on growth, 59
Escherichia coli, ethanol formation by, 94–5
— — genetic exchange in, 33–5
— — growth curve, 39
— — identification, 125–7
— — nitrite formation, 112
— — nutritional requirements, 46
— — size, 3
— — synchronous growth, 41
— — transamination, 105
Ethanol, conversion from pyruvic acid by yeast, 94
— formation by *E. coli*, 94–5
Eubacteriales, general properties, 121–42
Eucaryotes, 117–18
Eucaryotic cells, 117–18

Fats, bacterial metabolism of, 113–14
Fermentation, acetone-butanol, 95–6
— lactic, 95
Ferredoxin, 85, 88, 89
Fimbriae, 15–16
Flagella, 15–16
Fumagillin, 152
Fungi, 149–54
— chemical structure, 151–2
— classification, 150–1
— form, 150
— growth, 151
— metabolism, 152
— pathogenic, 154
— pharmaceutical importance of, 152–4
Fungi imperfecti, 150

Genetic exchange in bacteria, 33–6
Gibberella fujikuroi, gibberellic acid from, 153
Glucose, anaerobic catabolism of, 91–8
Glutamic acid, 45, 46
Glutamine, 45, 46
Glycine, 45, 46
Glycolysis, anaerobic, 91–8
Gram stain, 29–31
Gram-negative bacteria, toxins from, 26–7
Griseofulvin, 152

Growth, bacterial, *see* Bacteria, growth
— factors, 44–8

Haemophilus, general properties, 128–9
Hapten, 27
Heat, lethal effect, mechanism of, 62
— sterilising agent, 61
Histidine, 45–6
Hyaluronidase, 26
Hydrogen, formation by bacteria, 97–8
— utilisation by bacteria, 44
Hydrogen sulphide, formation from sulphur-containing amino acids, 106–7
Hydroxyproline, 46

IMViC tests, 126–7
Indole, formation of, 104–5
Interferon, 148–9
Iron, non-haem, 84
Irradiation, effect on bacterial growth, 65–7
Isoleucine, 45, 46

Klebsiella, general properties, 123
Krebs' citric acid cycle, 98–100

Lactic fermentation, 95
Lactobacillaceae, general properties, 131
Lactobacillus, general properties, 134
— nutritional requirements, 46
Lecithinases, 26
Leptospira, general properties, 142
Leucine, 45, 46
Leucocidins, 26
Leuconostoc, 3
— general properties, 133–4
— sucrose metabolism, 18
Luciferase, 24
Luminosity, bacterial, 24
Lysine, 45, 46

MacConkey's medium, 125–6
McIntosh and Fildes jar, 52
Mean generation time, 42–3
Meat, cooked, as a growth medium, 53
Membrane, cytoplasmic, 10–13
Mesophilic bacteria, 60
Mesosomes, 12
Metabolic studies, chemicals used in, 115
Metabolism, bacterial, *see* Bacteria, metabolism
Metabolite analogues, 69
Metachromatic granules, *see* Volutin granules
Methionine, 45, 46
Michaelis constant, 77, 79–80
Micro-aerophilic organisms, 51
Micrococcaceae, general properties, 129–31
Micrococcus, degradation of nucleic acids, 111
— general properties, 129
Microsporum, 154
Multiplication, bacterial, pattern of, 38–43
Muramic acid, 6–10
Mycobacteriaceae, general properties, 139–40
Mycobacterium, acid-fastness, 31
— general properties, 139–40

NAD-linked dehydrogenation, 84, 88–9
NADP-linked dehydrogenation, 84, 88–9
Neisseria, general properties, 131
— *catarrhalis*, toxins, 27
— *gonorrhoeae*, 2
— — toxins, 27
— *meningitidis*, nutritional requirements, 46
Neisseriaceae, general properties, 131
Neomycin, 141
Nicotinamide, 47
Nicotinic acid, 46, 47
Nitrate reduction, 112
Nitrogen, utilisation by bacteria, 44
Nitrogen-containing compounds, bacterial metabolism of, 101–8
Nocardiacea, 141

Nocardia, general properties of, 141
Nocardia luride, 141
Non-haem iron, 84
Novobiocin, 141
Nucleic acids, bacterial metabolism of, 108–11
Nutritional requirements of bacteria, 43–8
Nystatin, 141

Osmotic pressure, effect on bacterial growth, 63–5
Oxidation-reduction potentials, and anaerobic growth, 54–7
— — — bacterial metabolism, 83–7
Oxidative deamination, 104
Oxidisation, in bacterial cell, 82–7
Oxygen, bacterial requirements of, 50–1
— limitation and sporulation in *B. megaterium*, 20
Oxytetracycline, 141

Panthothenic acid, 46, 47
Pasteur effect, 87
Pasteurella, general properties, 127–8
Penicillin, 152
Penicillium, antibiotics from, 152
— *notatum*, 152
Peptide synthesis, bacterial, 103
Peptidoglycan, 6–10
Permeases, 13
pH, effect on bacterial enzyme action, 77
— — on bacterial growth, 57–9
Phenylalanine, 35, 36
Phosphorus, utilisation by bacteria, 44
Photoreactivation, 66–7
Photosynthetic autotrophic bacteria, 48
Physomycetes, 150
Pigments, bacterial, 23
Pili, 16
Pimelic acid, 46, 47
Plasmolysis, 10
Poliomyelitis virus, 146, 148
Procaryotes, 13, 117, 118
Procaryotic cells, 13, 117, 118
Proline, 46, 47
Protein breakdown, bacterial, 101–2
— synthesis, bacterial, 103
Proteinases, 26
Proteus, general properties, 123–4
— nutritional requirements, 46
— *vulgaris*, hydrogen sulphide production, 106–7
Protoplast, 11
Pseudomonadales, general properties, 119–20
Pseudomonas, general properties, 121
— *aeruginosa*, cultivation, 49
— — transamination, 105
Psychrophilic bacteria, 60
Pteroic acid, 46, 47
Pteroyglutamic acid, 46, 47
Purine, 46, 111
Pyridoxal, 46, 47
Pyrimidine, 46, 111
Pyrogallol, in cultivation of anaerobes, 51–2
Pyruvic acid, conversion to ethanol by yeast, 94
— — decomposition of by coliform organisms, 98
— — — by micro-organisms, 93

Redox potentials, *see* Oxidation-reduction potentials
Reproduction, bacterial, 31–6
Rhizopus nigricans, 153
Riboflavin, 46, 47
Ribose, 108
Ribosomes, 4, 14
Rickettsiae, general properties, 143
Rideal–Walker test, 124
Ristocetin, 141
RNA, 103, 108–9
Robertson's cooked meat medium, 53

Salmonella, general properties, 124–5
— toxins, 26–7
— *schottmülleri* (*paratyphi*), 2, 124, 126
— *typhosa*, identification with bismuth sulphite agar, 126
— — nutritional requirements, 46
Serine, 45, 46
Serratia, general properties, 123

Serratia, marcescens, cultivation, 49
— — pigment, 23
Shigella, general properties, 125
— identification with bismuth sulphite agar, 126
— toxins, 26–7
Spheroplasts, 11
Spiramycin, 141
Spirochaetales, general properties, 141–2
Spores, bacterial, 18–23
Staining, of bacteria, 28–31
Staphylococcus, general properties, 129–31
— teichoic acid in cell wall, 8–9
— tests for, 130
— toxins, 26
— *aureus*, 130
— — cell wall, 5, 6–7
— — nutritional requirements, 46
— — pigment, 23
— — size, 3
— *citreus*, pigment, 23
Sterilisation, by heat, 61
Steroids, synthesis, by fungi, 153–4
Stickland reaction, 106
Streptococcus, general properties, 132–3
Streptococcus, nutritional requirements, 46
— toxin, 26
— *viridans*, 2
Streptodornase, 25
Streptokinase, 25
Streptomyces, general properties, 140–1
— antibiotics produced by, 141
Streptomycin, 141
Substrate concentration, effect on bacterial enzyme action, 77–8, 80
Sulphur, utilisation by bacteria, 44

Teichoic acids, in bacterial cell walls, 8–9
Temperature, effect on bacterial enzyme action, 79
— — on bacterial growth, 59 63
Tetracycline, 141
Thermal death rate, 61–2
Thermal death rate time, 61–2
Thermophilic bacteria, 60
Thiamine, 46
Thioglycollic acid, in cultivation of anaerobes, 53
Threonine, 46
Toxins, bacterial, 24–7
Trace elements, in bacterial growth, 44, 48
Transamination, 105
Transduction, 35–6
Transformation, 35
Transport, through membranes, 13
Treponema, general properties, 125–6
Tricarboxylic acid cycle, 85–7, 99–100
Trichophyton, 138
Tryptophan, 45, 46
Tryptophanase, 92
Tyrosine, 45, 46

Ubiquinone, 84
Uracil, 46

Valine, 35, 36
Vibrio, general properties, 121
Viomycin, 141
Viruses, 143–8
— biological properties of, 145–7
— chemical nature of, 145
— cultivation of, 147–8
— diseases caused by, 147
— interferon, 148
— reaction to physical environment, 147
— shape and size, 145
Voges–Proskauer reaction, 97, 127
Volutin granules, 14

Walls, of bacterial cells, 3–10
— Gram-negative cells, 9–10
— — positive cells, 8–9
— spores, 18–20
Water activity, 64–5
Water, for growth, 38, 64–5

Yeast, conversion of pyruvic acid to ethanol by, 94

Ziehl–Neelsen method of staining, 30–1